CALCULATION EXAMPLES

ARITHMETIC 2

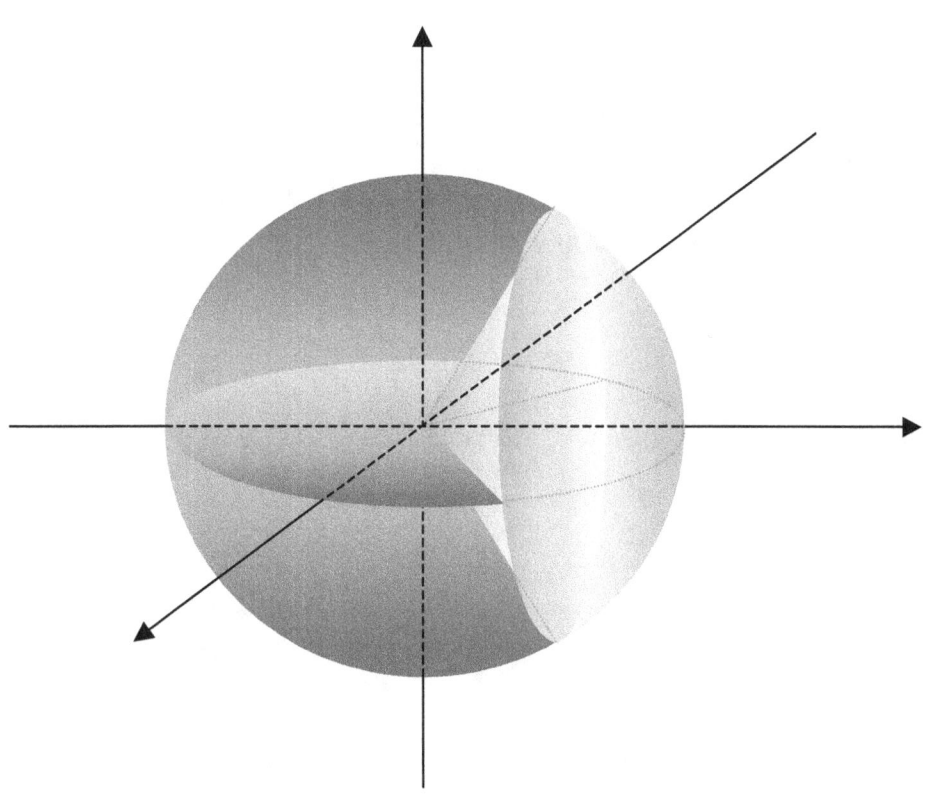

Seong R. KIM

Dear students:

Students need the best teacher, so you need examples, because examples are the best teacher. All the examples here are fully worked, and explain **how** the basic and essential tools in math are made, together with **what** they are, **how** they work, and **how** to work with them. Such tools include numbers, formulas, identities, equations, laws, etc.

Examples here begin with easy ones, of course. Covering every meter and yard properly, we can cover thousands of miles and kilometers. And it is particularly the case in math.

Of those examples therefore, some might even look too easy for you. It's not that easy though, to come up with those examples. Anyways, the bigger and the taller the tree, the deeper and the stronger the root.

Doing math, we work with ideas and run ideas, because every thing in math is an idea. A number is an idea, for instance, and the same is true for a line or circle, too. And putting ideas together, we build another, which becomes the base or an element of another, and each is connected. And that's the way your math grows. So you get to build a circuit, and sometimes, need to fill the gap or repair the circuit so that you get the sense of it.

So your calculation runs properly, and you get the problem solved.

The examples have been made and arranged so that they get tougher (or sometimes easier for some reason) as you proceed with them. In particular, similar examples with some variations are strategically repeated so that you can get the ideas or the tools tricky or complicated, and can get them mastered.

This book is however, nothing but a bunch of examples until you get it powered. How then, to get it powered, and make it run and work for you?

Just read it, and then, do each example in writing. And it is important to note that you do it in **your** writing. Just watching someone doing it, you just only feel that you can do it. If you do it, you can do it, but if you don't, we can hardly. It's a cliché, of course, but is always true that knowing is one thing and doing is another.

I've been helping students grow, take care of, and run their own math. The area covers algebra and geometry for high school or college students, and is especially for equations (for unknowns or curves), functions, and their graphs, which are the basic elements in calculus, which's been the core of my interest from my early age in high school.

Of my students, some are quite poor in math, and thus, are afraid of or hate math, some require special education because of exceptional intelligence, some are smart enough, some are naïve and diligent, some are clever but lazy, and most behave in general. All the students are badly after though, one thing in common: a strong and secure math skill. It is of course, the prime objective of my work, and I'm always happy to and eager to help them achieve it. The problem was however, that many of them wanted it to be purchased. And the question is, can we buy it?

We can buy the means, of course. And a solid math skill is feasible, too. We know however, we can't buy love, and the same is true for the math skill, too. It's not what we can buy or sell, and not what we can give or take. It is however, what we can grow, and need to grow. Your math grows as much as you grow and take care of it. So does mine.

What math then, do students most often do or use in high schools or colleges?

It is algebra and geometry. What algebra though?

Elementary algebra, of course
Doing the algebra, we work with numbers (many in kinds), constants, variables, ratios, rates, expressions, equations, inequalities, functions, identities, formulas, laws, etc., together with signs and symbols. And if we want to do algebra properly, we want to know their natures and how they mingle with each other.

So studying math ideas or tools, you want to know **what** they are, **how** they work, and **how** to work with them or **what** to do with them.   What then, about the geometry?

Basically, the geometry has much to do with shapes, positions, and angles. The shapes begin with triangles and circles, and move on to rectangles, squares, parallelograms or rhombuses, trapezoids, tetragons, other polygons, polyhedrons, etc.

Doing the geometry, too, though, we need to do the algebra stated above. So it is analytic geometry, often called coordinate geometry, too. And doing it, we can specify positions using coordinates. So in the geometry, basically, we work with graphs. Putting a math idea in a graph, we can not only effectively think about it but actually see it, too, and therefore, can efficiently work with it. What idea then, is it?

The idea begins with a point, line, parabola, circle, ellipse, and hyperbola, called a conic section or basic curve, and then, moves on to other curves, planes, surfaces, volumes, and other objects in various dimensional spaces, together with vectors.

And using an angle, we can specify an amount of turn or change in direction.

So learning, using, or applying those ideas or math tools, we get to solve problems.

And this book can help. It can help learn them, and use them so that you can navigate to find solutions to problems. And in particular, it can help come up with answers to those **what**s and **how**s stated above. So it can help you grow and run your own math, and thus, can help achieve your solid math skill.

It is however, not a magic book giving you a math skill of high caliber overnight. And it can have many mistakes, too. There is no magic, and math is full of facts and ideas. And it is after all, not me and not your teacher but you who put together some of those facts and ideas, and understand it. Putting facts and ideas together, understanding it, and taking care of what you have learned, you grow your math. And this book can help.

This is a book of examples designed to help you grow your math, and assumes that you are a real beginner. This book requires though, time and effort, the amount of which need to be substantial, too, but will be worth it. That's because you want a substantial achievement, and will get it. And probably, you will get to see this book helping you get there much faster than expected. And then, you will get to see the way math runs.

In math, everything is an idea. So is a problem. And solving it, we put it many different ways. For instance, while expanding or reducing it, or modifying or converting it, we keep searching for the solution, approaching the solution, and eventually, can get there. So don't look for the solution outside the problem. The solution is inside the problem if the problem is properly made.

If it is not, no solution is the solution. And in fact, it is often the case a problem itself is the solution. We can put a problem in many different ways, and eventually, can end up with the solution.   How come then, is the solution no other than the problem?

For instance, the solution to $3232 \div 101$ is 32.   And we can put it this way:

$$3232 \div 101 = \frac{3232}{101} = \frac{32 \times 101}{101} = \frac{32}{1} = 32 \implies 3232 \div 101 = 32.$$

And we can get this, too: $32 \implies 3232 \div 101$.   How?

$$32 = \frac{32}{1} = \frac{32 \times 101}{101} = \frac{3232}{101} = 3232/101 = 3232 \div 101.$$ Too easy?

For another instance, the solution to $ax^2 + bx + c = 0$ is: $x = \frac{-b \pm \sqrt{b^2 - 4ac}}{2a}$, which is called the quadratic formula.   How come then, is the solution no other than the problem?

We can put it this way:

$$x = \frac{-b \pm \sqrt{b^2 - 4ac}}{2a} \implies 2ax = -b \pm \sqrt{b^2 - 4ac} \implies 2ax + b = \pm \sqrt{b^2 - 4ac}$$

$$\implies (2ax + b)^2 = b^2 - 4ac \implies 4a^2x^2 + 4abx + b^2 = b^2 - 4ac$$

$$\implies 4a^2x^2 + 4abx = -4ac \implies ax^2 + bx = -c \implies ax^2 + bx + c = 0.$$

And we can get this, too: $ax^2 + bx + c = 0 \implies x = \frac{-b \pm \sqrt{b^2 - 4ac}}{2a}$.   How?

$$ax^2 + bx + c = a(x^2 + \tfrac{b}{a}x) + c = a(x^2 + \tfrac{b}{a}x + \tfrac{b^2}{4a^2} - \tfrac{b^2}{4a^2}) + c = a(x^2 + \tfrac{b}{a}x + \tfrac{b^2}{4a^2}) - \tfrac{b^2}{4a} + c$$

$$= a(x + \tfrac{b}{2a})^2 - \tfrac{b^2 - 4ac}{4a} = 0 \implies a(x + \tfrac{b}{2a})^2 = \tfrac{b^2 - 4ac}{4a} \implies (x + \tfrac{b}{2a})^2 = \tfrac{b^2 - 4ac}{4a^2} \implies x + \tfrac{b}{2a} = \pm\sqrt{\tfrac{b^2 - 4ac}{4a^2}}$$

$$\implies x = -\tfrac{b}{2a} \pm \tfrac{\sqrt{b^2 - 4ac}}{2a} = \tfrac{-b \pm \sqrt{b^2 - 4ac}}{2a} \implies x = \tfrac{-b \pm \sqrt{b^2 - 4ac}}{2a}.$$

And we call the set of processes above, algebra.

So if a problem is well defined, that is, if it makes sense, we should be able to get it solved the way below:

A problem $\Rightarrow$ ... $\Rightarrow$ ... $\Rightarrow$ the solution, and thus: **the problem $\Rightarrow$ the solution**.

So solving a problem, we put it many different ways so that we can get to the solution.

And that's the way, math runs.

May your math run very well.

Seong R. Kim

B.S. Math. Michigan Tech. Univ.   M.S. Math. Rensselaer Polytechnic Institute

Notes:

This book is one of three books about some basics in numbers, and covers arithmetic operations. And the operations are on numbers called real numbers. We can classify real numbers in two ways. In one, a number is an integer or a non-integer, which is not an integer. And in the other, a number is rational or irrational, which is not rational.

So if a number is not irrational, it is rational. Among rational numbers, we have integers and non-integers. And if a number is not an integer, it is a non-integer. Among non-integers, we have rational numbers and irrational numbers. It's confusing, isn't it?

So usually, we classify real numbers into three groups: integers, rational numbers, and irrational numbers. And this book is for your study of math using integers.

The book is designed for those students who want to study word problems, where students need to manage integers and set up expressions. And you will get to see some nature of integers and some tools useful or handy using or working with integers. The tools are math ideas, called theorems or formulas, which you can use doing problems with multiples or divisors, etc.

What then, about the other numbers?

They are covered in **Calculation Examples Arithmetic 3**, which covers thus, numbers rational as 0.5, -3/2, or 5/3, and numbers irrational as a square root of 2.

What then, about **Calculation Examples Arithmetic 1**?

The book covers integers, and helps understand what integers are about and how they mingle with arithmetic operations so that you can not only use integers properly but can develop your own idea to make use of them, too, solving problems, of course.

It doesn't just cover though, how to do arithmetic with integers, that is, how to do additions, subtractions, multiplications, and divisions. But it helps also understand the nature of integers and the nature of arithmetic operations as additions or divisions. And in particular, it covers the nature and the use of numbers called inverses. What inverse?

It can be called the negative or the reciprocal depending on the arithmetic operation. So this book covers those math ideas (concepts), and helps you understand them, and use them efficiently as well as properly. It can help speed up your calculation, too.

And thus, all the three books will help you grab math ideas often used in real life as well as in math courses. The ideas are about real numbers, their arithmetic operations, and their nature, so you will get to see what those numbers are about and how they mingle with arithmetic operations, and will get to develop your own idea to make use of those, solving problems, of course.

In short, the books help you strengthen fundamentals in math to increase your skill of algebra, that is, calculation techniques, providing you with examples, showing all the steps and the ideas behind, and explaining what the math ideas are about.

Contents

Examples L

0.   Suppose we make integers using numbers from a set A = {**7, 0, 9, 8**}, and making each integer, we have to use three different numbers in the set.
What are then, all the integers less than 700?

1.   Suppose we make integers using numbers from a set B = {**5, 0, 9, 4, 6**}, and making each integer, we have to use four different numbers in the set.
Find the number of all the integers < 5000.

2.   Doing examples below, use numbers in a set C = {**3, 5, 0, 7**}. Note however, each number in the set can be used once only. That is, no number can be used more than once.

2.0.   What is the smallest four-digit integer?

2.1.   What is the second smallest four-digit integer?

2.2.   What is the largest four-digit integer?

2.3.   What is the second largest four-digit integer?

3.   Find all the integers described as follows:

   A. 1's digit is smaller than 1000's digit.
   B. 10's digit is lager than 1000's digit.
   C. 1000's digit is twice 100's digit.
   D. Each integer is between 6000 and 7000.
   E. Adding together all the numbers in all the digits in each integer, we get 19.

Suggestions or Solutions
To the Problems in the Examples L

0. **Suppose we make integers using numbers from a set $A = \{7, 0, 9, 8\}$, and making each integer, we have to use three different numbers in the set. What are then, all the integers less than 700?**

They are: 98, 97, 89, 87, 79, and 78.

If not sure of how to get it, follow the steps below:

If a number is less than 700, the 100's digit is less than or equal to 6.
What number then, from the set, can we use for the 100's digit in an integer we make?

We do not have to make a 3-digit integer, because the problem is not asking 3-digit integers. What do we mean by though, a 3-digit integer?

A 3-digit integer is between 1000 and 99. And in math, assuming such an integer is N, we can put it this way: $99 < N < 1000$.

More precisely though, it is greater than or equal to 100 and less than or equal to 999. And in math, we can put it this way: $100 \leq N \leq 999$.

And more specifically, since N is an integer, we can put it the way below, too:
$N = 100, 101, 102, \ldots 998, 999$.

So in this problem, making an integer, we have only to use three numbers from the set. And thus, we can use 0 for the 100's digit in an integer we make.

So we can make: 098, 097, 089, 087, 079, and 078, which are all the integers less than 700, and of course, we usually put them this way: 98, 97, 89, 87, 79, and 78.

1.   Suppose we make integers using numbers from a set $B = \{5, 0, 9, 4, 6\}$, and making each integer, we have to use four different numbers in the set. Find the number of all the integers < 5000.

There are 48 integers.

If not sure of how to get it, follow the steps below:

If a number is less than 5000, what can be the 1000's digit in the number?

It can be less than or equal to 4.
So assuming N is the integer we want to make, we can put it the way below:
$N = 4000 + 100a + 10b + c$, where a, b, and c are numbers that belong to $C = \{5, 0, 9, 6\}$.

So we now have to choose a number for each of a, b, and c.
Then, to begin with, how many choices do we have for a?

We have four. That's because any of 5, 0, 9, and 6 can be used as a. What then, about b?

Assuming we have chosen a number for a, we have three choices for b. That's because, of the four numbers in the set C, one has been used as a. What then, about c?

Assuming we have chosen two numbers for a and b, we have two choices for c. That's because, of the four numbers in the set C, one has been used as a, and another has been used as b.

So we have 4 choices for a. And for each choice for a, we have three choices for b. And for each choice for b, we have two choices for c.

And thus, the number of all the integers < **5000** is: 4 x 3 x 2 = 24.

And the integers less than 5000 are as follows:

4965, 4960, 4956, 4950, 4906, 4905

4695, 4690, 4659, 4650, 4609, 4605

4596, 4590, 4569, 4560, 4509, 4506

4096, 4095, 4069, 4065, 4059, 4056

Is that it though?

We do not have to make a 4-digit number, and making a number, we have only to use four different numbers from the set.

That is to say that we can begin with putting 0 in the 1000's digit when making an integer < **5000**.

So assuming M is such an integer, we can put it the way below:

$M = 1000 \times 0 + 100a + 10b + c$, where a, b, and c belong to $D = \{5, 9, 4, 6\}$.

That is, a indicates 100's digit, b indicates 10's digit, and c indicates 1's digit.

So we have 4 choices for a. And for each choice for a, we have three choices for b. And for each choice for b, we have two choices for c.

And thus, in this case, too, the number of all the integers < **5000** is: $4 \times 3 \times 2 = 24$. And those integers are as follows:

0965, 0964, 0956, 0954, 0946, 0945

0695, 0694, 0659, 0654, 0649, 0645

0596, 0594, 0569, 0564, 0549, 0546

0496, 0495, 0469, 0465, 0459, 0456

And thus, the number of all the integers < **5000** is 48.

2.   Doing examples below, use numbers in a set *E* = {3, 5, 0, 7}. Note however, each number in the set can be used once only. So no number can be used more than once.

2.0.   What is the smallest four-digit integer?

Making a 4-digit integer, what digit cannot be 0?

It is the highest digit, which is the 1000's digit. So we cannot use 0 in the 1000's digit since we need to make a four-digit integer.

We can however, put 0 in the second highest digit, which is in this case, the 100's digit. And the smaller the number is in a particular digit, the smaller the integer is.
So the smallest integer is: 3057.

2.1.   What is the second smallest four-digit integer?

The second smallest is larger than the smallest, and is the smallest of all the others.
So the second smallest is: 3075.

2.2.   What is the largest four-digit integer?

The larger the number is in a particular digit, the larger the integer is.
So the largest integer is: 7530.

2.3.   What is the second largest four-digit integer?

The second largest is smaller than the largest, and is the largest of all the others.
So the second largest is: 7503.

3. **Find all the integers described as follows:**

    A. **1's digit is smaller than 1000's digit.**

    B. **10's digit is lager than 1000's digit.**

    C. **1000's digit is twice 100's digit.**

    D. **Each integer is between 6000 and 7000.**

    E. **Adding together all the numbers in all the digits in each integer, we get 19.**

They are: 6373, 6382, and 6391.

If not sure of how to get it, follow the steps below:

To begin with, each integer we make is between 6000 and 7000.
So assuming N is the integer, we get: $\mathbf{6000 < N < 7000}$.

Thus next, we can see that N is a 4-digit integer positive, of course.
So we can set: $N = \boldsymbol{abcd}$, where $\boldsymbol{a}$ is the number in the 1000's digit, $\boldsymbol{b}$ is the one in the 100's digit, $\boldsymbol{c}$ is the one in the 10's digit, and $\boldsymbol{d}$ is the one in the 1's digit.

Thus, note that in this case:

N is $\boldsymbol{a} \times \mathbf{1000} + \boldsymbol{b} \times \mathbf{100} + \boldsymbol{c} \times \mathbf{10} + \boldsymbol{d}$, and thus, is <u>not</u> $\boldsymbol{a} \times \boldsymbol{b} \times \boldsymbol{c} \times \boldsymbol{d}$.

Normally however, just saying $N = \boldsymbol{abcd}$, we mean: $N = \boldsymbol{a} \times \boldsymbol{b} \times \boldsymbol{c} \times \boldsymbol{d}$. So for instance, if we are given just $\boldsymbol{K} = \boldsymbol{uv}$ with no more specification, it means: $\boldsymbol{K} = \boldsymbol{u} \times \boldsymbol{v}$.
Then first, in $N = \boldsymbol{abcd}$, what can be the number $\boldsymbol{a}$?

We know: $\mathbf{6000 < N < 7000}$, and N is an integer.
So we get: $\mathbf{6001 \leq N \leq 6999}$, that is, N can be any of the integers below:
$\mathbf{6001, 6002, \ldots, 6998}$, and $\mathbf{6999}$. And we have: $N = \boldsymbol{abcd}$. So we get: $\boldsymbol{a = 6}$.

Next, the statement **C** says: 1000's digit is twice 100's digit.

And we know a is in the 1000's digit, and b is in the 100's digit.

So we get: $a = 2 \times b$.

And thus, we get: $b = 3$ since $a = 6$.

So we now have: $N = 63cd$.

Next, the statement **B** says: 10's digit is lager than 1000's digit.

So we get: $c > 6 \Rightarrow c = 7, 8$, or 9.

Next, the statement **A** says: 1's digit is smaller than 1000's digit.

So we get: $0 \leq d < a = 6 \Rightarrow d = 0, 1, 2, 3, 4$, or 5.

And next, from the statement **E**, we get: $6 + 3 + c + d = 19 \Rightarrow c + d = 10$.

So what can be the next?

We have:

$c = 7, 8$, or 9.

$d = 0, 1, 2, 3, 4$, or 5.

And $c + d = 10$.

So we can pair up the numbers for c and d the way below:

$(c, d) = (7, 3), (8, 2)$, and $(9, 1)$. What do we mean by though, $(c, d) = (7, 3)$?

If $c = 7$, then $d = 3$.

And we have: $N = 63cd$.

So the solutions is: 6373, 6382, and 6391.

8

Examples M

0.0.   Suppose $A78 + B5C = DEE4$, where A, B, C, D, and E indicate digits.
So for instance, $A78 = A \times 100 + 7 \times 10 + 8$. What then, is the sum: $A + B + C + D + E$?

0.1.   Assuming: $38A7 + 297B + C789 = 8D52$, find the sum: $A + B + C + D$.

1.   Using the four numbers below, make two different 2-digit integers so that the product of the two is the largest in each of the two cases below.

1.0.   Using at least once each of the four numbers
1.1.   Using once only each of the four numbers

2, 5, 7, and 8

2.   Clark made two 2-digit integers using numbers as follows: 1, 6, 8, and 9.
Then, he flipped both integers. For instance, flipping 91, he gets 19. And then, he took the sum of the two integers, and got 87. What integers then, did he flip? That is, find the two 2-digit integers he made in the first place.

3.   Suppose A and B are two different 1-digit integers. What then, are AB and A if we get: $AB \times A = 280$, where A and B indicate digits? For example, $AB = A \times 10 + B$.

4.   Romans in the past used numbers the way below, and some of them are still used now. They are often used for decorative purposes.

I = 1,   II = 2,   III = 3,   IV = 4,   V = 5,   VI = 6,   VII = 7,   VIII = 8,   IX = 9,

X = 10,   XL = 40,   L = 50,   XC = 90,

C = 100,   CD = 400,   D = 500,   CM = 900,   M = 1000

For instance:

XXXIV = X + X + X + IV = 10 + 10 + 10 + 4 = 34

MCMXCII = M + CM + XC + II = 1000 + 900 + 90 + 2 = 1992

4.0   Put the Roman numbers below in integers.

XLVI, CMLV, MCMLXXXIX, and MMMDCLXXVIII

4.1.   Express the following numbers in the Roman numbers.

94, 952, 1697, and 2069.

Suggestions or Solutions
To the Problems in the Examples M

0.0.   Suppose $A78 + B5C = DEE4$, where $A, B, C, D,$ and E indicate digits. So for instance, $A78 = A$ x $100 + 7$ x $10 + 8$. What then, is the sum: $A + B + C + D + E$?

It is 11 or 22.

If not sure of how to get it, follow the steps below:

To begin with, we can put the three numbers the way below:

$A78 = A$ x $100 + 7$ x $10 + 8$ x 1

$B5C = B$ x $100 + 5$ x $10 + C$ x 1

$DEE4 = D$ x $1000 + E$ x $100 + E$ x $10 + 4$ x 1

So we get: $A78 + B5C = (A + B)$ x $100 + (7 + 5)$ x $10 + (8 + C)$ x 1

$= (A + B)$ x $100 + (10 + 2)$ x $10 + (8 + C)$ x 1

$= (A + B + 1)$ x $100 + 2$ x $10 + (8 + C)$ x $1.$

And thus, we can set:

$(A + B + 1)$ x $100 + 2$ x $10 + (8 + C)$ x $1 = D$ x $1000 + E$ x $100 + E$ x $10 + 4$ x 1

So first, comparing the 1's digits in both sides, we get: $8 + C = 4$, which is however, not possible, because we have: $0 \leq C \leq 9$, and C is an integer.

And thus, we need to get: $8 + C = 14$, so we get: $C = 6$. So we need to set:

$(A + B + 1)$ x $100 + (2 + 1)$ x $10 + 4$ x $1 = D$ x $1000 + E$ x $100 + E$ x $10 + 4$ x $1.$

So next, comparing the 10's digits in both sides, we get: $E = 3$. And thus, we can set:

$(A + B + 1)$ x $100 + 3$ x $10 + 4$ x $1 = D$ x $1000 + 3$ x $100 + 3$ x $10 + 4$ x $1.$

So next, comparing the 100's digits in both sides, we get: $A + B + 1 = 3$.

And we know: $0 \leq A \leq 9$, $0 \leq B \leq 9$, and A and B are integers. So we can have three cases as follows: $(A, B) = (0, 2)$, $(1, 1)$, or $(2, 0)$. Anyway, we get: $A + B = 2$.

Isn't it the case though, where $A78$ and $B5C$ are 3-digit integers?
That is, why not: $1 \leq A \leq 9$, and $1 \leq B \leq 9$, but: $0 \leq A \leq 9$, and $0 \leq B \leq 9$?

It is not said that $A78$ and $B5C$ are 3-digit integers. So A and B can be 0.
And it is not said either that $DEE4$ is a 4-digit integer.
So it can be the case where we get: $D = 0$.

Now, we have: $A + B = 2$.
So we can now set: $3 \times 100 + 3 \times 10 + 4 \times 1 = D \times 1000 + 3 \times 100 + 3 \times 10 + 4 \times 1$.
Thus, next, comparing the 1000's digits in both sides, we get: $D = 0$.

What about a case where $A + B + 1 = 13$ though?

We can get, of course: $A + B = 12$, too. For instance, we can get: $A = 3$, and $B = 9$.
And in that case, we need to set:

$1 \times 1000 + 3 \times 100 + 3 \times 10 + 4 \times 1 = D \times 1000 + 3 \times 100 + 3 \times 10 + 4 \times 1$.

So we get: $D = 1$. And thus, we get:

$(A + B) + C + D + E = 2 + 6 + 0 + 3 = 11$.

$(A + B) + C + D + E = 12 + 6 + 1 + 3 = 22$.

0.1.   Assuming: $38A7 + 297B + C789 = 8D52$, find the sum: $A + B + C + D$.

To begin with, we can put the four numbers the way below:

$38A7 = 3 \times 1000 + 8 \times 100 + A \times 10 + 7$

$297B = 2 \times 1000 + 9 \times 100 + 7 \times 10 + B$

$C789 = C \times 1000 + 7 \times 100 + 8 \times 10 + 9$

$8D52 = 8 \times 1000 + D \times 100 + 5 \times 10 + 2$

So we get: **$38A7 + 297B + C789$**

$= (3 + 2 + C) \times 1000 + (8 + 9 + 7) \times 100 + (A + 7 + 8) \times 10 + (7 + B + 9)$

$= (C + 5) \times 1000 + (20 + 4) \times 100 + (A + 10 + 5) \times 10 + (B + 10 + 6)$

$= (C + 5 + 2) \times 1000 + (4 + 1) \times 100 + (A + 5 + 1) \times 10 + (B + 6)$

$= (C + 7) \times 1000 + 5 \times 100 + (A + 6) \times 10 + (B + 6)$

Thus, we get: **$38A7 + 297B + C789 = 8D52$** $\Rightarrow$

$(C + 7) \times 1000 + 5 \times 100 + (A + 6) \times 10 + (B + 6) = 8 \times 1000 + D \times 100 + 5 \times 10 + 2$.

So first, comparing the 1's digits in both sides, we get: $B + 6 = 2$, which is however, not possible, because we have: $0 \leq B \leq 9$, and B is an integer.

And thus, we need to get: $B + 6 = 12$, so we get: $B = 6$. And thus, we need to set:

$(C + 7) \times 1000 + 5 \times 100 + (A + 6 + 1) \times 10 + 2 = 8 \times 1000 + D \times 100 + 5 \times 10 + 2$.

So next, comparing the 10's digits in both sides, we get: $A + 6 + 1 = A + 7 = 5$, which is however, not possible, because we have: $0 \leq A \leq 9$, and A is an integer.

And thus, we need to get: $A + 7 = 15$, so we get: $A = 8$. So we need to set:

$(C + 7) \times 1000 + (5 + 1) \times 100 + 5 \times 10 + 2 = 8 \times 1000 + D \times 100 + 5 \times 10 + 2$.

So next, comparing the 100's digits in both sides, we get: $D = 6$. And thus, we can set:

$(C + 7) \times 1000 + 6 \times 100 + 5 \times 10 + 2 = 8 \times 1000 + 6 \times 100 + 5 \times 10 + 2$.

And next, comparing the 1000's digits in both sides, we get: $C + 7 = 8 \Rightarrow C = 1$.

And thus, we get: $A + B + C + D = 8 + 6 + 1 + 6 = 21$.

14

1. Using the four numbers below, make two different 2-digit integers so that the product of the two is the largest in each of the two cases below.

1.0. Each number can be used more than once.

1.1. Each number can be used once only.

2, 5, 7, and 8

> The larger the numbers are, the larger their product gets.
> Basically, multiplying two numbers, we do it digit by digit.
> And the larger the digit, the larger the number.

1.0. Making first, all possible 2-digit integers, we can get:

22, 25, 27, 28, 55, 52, 57, 58, 77, 72, 75, 78, 88, 82, 85, and 87.

And next, taking from the list above, the largest and the second largest, and taking the product of the two, we get the largest product. And the two are 88 and 87.

1.1. Making first, all possible pairs of two different 2-digit integers, we can get:

(25 and 78), (25 and 87), (52 and 78), (52 and 87), (27 and 58), (27 and 85), (72 and 58), (72 and 85), (28 and 57), (28 and 75), (82 and 57), and (82 and 75)

And among all the pairs above, either of (72 and 85) and (82 and 75) makes the product the largest. And taking the product of the two in each pair, we get:
85 x 72 = 6120, and 82 x 75 = 6150.

So the two integers we have to make are 82 and 75.

2.   Clark made two 2-digit integers using numbers as follows: 1, 6, 8, and 9. Then, he flipped both integers. For instance, flipping 91, he gets 19. And then, he took the sum of the two integers, and got 87. What integers then, did he flip? That is, find the two 2-digit integers he made in the first place.

The two integers made in the first place can be: **(81, 69)** or **(86, 91)**.

If not sure of how to get it, follow the steps below:

Suppose first, *AB* and *CD* are the two integers made in the first place.
Note that *AB* is not *A* x *B* but *A* x **10** + *B*. And the same is true for *CD*, too.

Then, *A* and *C* are numbers in the 10's digits, and *B* and *D* are numbers in the 1's digits. And of course, *A* is one of 1, 6, 8, and 9. And the same is true for *B*, *C*, and *D*, too.

Next, flipping the two integers, he gets: *BA* and *DC*.
Then, *B* and *D* are the numbers in the 10's digits, and *A* and *C* are the numbers in the 1's digits. And taking the sum of the two integers, he gets: *BA* + *DC* = 87.

What then, is *A* + *C*?

It can be the case where *A* + *C* = 7, or *A* + *C* = 17.
And we know that *A* and *C* can be any of the numbers in the list: 1, 6, 8, and 9.
So assuming first, *A* + *C* = 7, we get: (*A*, *C*) = (1, 6) or (6, 1).

And next, we have: *B* + *D* = 8, which is however, not possible, because *B* and *D* have to be from the list, but the list does not have the numbers for *B* and *D*.
So it cannot be the case where *A* + *C* = 7.

Thus next, moving on to the case where *A* + *C* = 17, we get: (*A*, *C*) = (8, 9) or (9, 8).
And we can put *BA* and *DC* this way: *BA* = *B* x **10** + *A*, and *DC* = *D* x **10** + *C*.

Thus, we get: $BA + DC = (B + D) \times 10 + A + C$.

And we know: $A + C = 17$, and $BA + DC = 87$.

So we get: $BA + DC = (B + D) \times 10 + 17 = (B + D + 1) \times 10 + 7 = 87$.

Thus next, comparing the 10's digits, we get: $B + D + 1 = 8 \Rightarrow B + D = 7$.

And we know that B and D can be any of the numbers in the list: 1, 6, 8, and 9.

So we get: $(B, D) = (1, 6)$ or $(6, 1)$.

And thus, putting threads together, we have:

$(A, B, C, D) = (8, 1, 9, 6)$, $(8, 6, 9, 1)$, $(9, 1, 8, 6)$, or $(9, 6, 8, 1)$.

So assuming first, $(A, B, C, D) = (8, 1, 9, 6)$, we get: $BA + DC = 18 + 69 = 87$.

Assuming next, $(A, B, C, D) = (8, 6, 9, 1)$, we get: $BA + DC = 68 + 19 = 87$, too.

Assuming next, $(A, B, C, D) = (9, 1, 8, 6)$, we get: $BA + DC = 19 + 68 = 87$, too.

And assuming next, $(A, B, C, D) = (9, 6, 8, 1)$, we get: $BA + DC = 69 + 18 = 87$, also.

So the two integers made in the first place seem to make either of the four pairs below:

$(AB, CD) = (81, 96)$, $(86, 91)$, $(91, 86)$, and $(96, 81)$.

However, $(81, 96)$ is no other than $(96, 81)$, and $(86, 91)$ is no other than $(91, 86)$.

And thus, the two integers made in the first place can be: $(81, 69)$ or $(86, 91)$.

3.   Suppose A and B are two different 1-digit integers. What then, are AB and A if we get: $AB \times A = 280$, where A and B indicate digits? For example, $AB = A \times 10 + B$.

   $AB = 56$, and $A = 5$.   *If not sure of how to get it, follow the steps below:*

To begin with, we can set: $AB \times A = (A \times 10 + B) \times A = A \times A \times 10 + A \times B$.
So we get: $10A^2 + A \times B = 280$, where $10A^2$ is 10 times A^2, that is, $10 \times A^2$.
And we know $10A^2$ is a multiple of 10. What then, about $A \times B$ in $10A^2 + A \times B = 280$?

We can say $A \times B$ is a multiple of 10, too, because 280 is a multiple of 10.
What then, A and B can be?

First, each of A and B can be one of 0, 1, 2, 3, … 8, and 9, since A and B are numbers used for digits.   So next, we can put (A, B) the way below:

$(A, B) = (2, 5), (5, 2), (4, 5), (5, 4), (6, 5), (5, 6), (8, 5),$ and $(5, 8)$.

And thus, taking the product $AB \times A$ using each pair above, we get:

$(A, B) = (2, 5) \Rightarrow (AB, A) = (25, 2) \Rightarrow AB \times A = 25 \times 2 \neq 280$.
$(A, B) = (5, 2) \Rightarrow (AB, A) = (52, 5) \Rightarrow AB \times A = 52 \times 5 \neq 280$.
$(A, B) = (4, 5) \Rightarrow (AB, A) = (45, 4) \Rightarrow AB \times A = 45 \times 4 \neq 280$.
$(A, B) = (5, 4) \Rightarrow (AB, A) = (54, 5) \Rightarrow AB \times A = 54 \times 5 \neq 280$.
$(A, B) = (6, 5) \Rightarrow (AB, A) = (65, 6) \Rightarrow AB \times A = 65 \times 6 \neq 280$.
• $(A, B) = (5, 6) \Rightarrow (AB, A) = (56, 5) \Rightarrow AB \times A = 56 \times 5 = 280$.
$(A, B) = (8, 5) \Rightarrow (AB, A) = (85, 8) \Rightarrow AB \times A = 85 \times 8 \neq 280$.
$(A, B) = (5, 8) \Rightarrow (AB, A) = (58, 5) \Rightarrow AB \times A = 58 \times 5 \neq 280$.

And thus, we get: $AB = 56$, and $A = 5$.

4. **Romans in the past used numbers the way below, and some of them are still used now. They are often used for decorative purposes.**

I = 1,   II = 2,   III = 3,   IV = 4,   V = 5,   VI = 6,   VII = 7,   VIII = 8,   IX = 9,
X = 10,   XL = 40,   L = 50,   XC = 90,
C = 100,   CD = 400,   D = 500,   CM = 900,   M = 1000

For instance: XXXIV = X + X + X + IV = 10 + 10 + 10 + 4 = 34
MCMXCII = M + CM + XC + II = 1000 + 900 + 90 + 2 = 1992

4.0 **Put in integers the Roman numbers below:**

XLVI, CMLV, MCMLXXXIX, and MMMDCLXXVIII

XLVI = XL + VI = 40 + 6 = 46

CMLV = CM + L + V = 900 + 50 + 5 = 955

MCMLXXXIX
= M + CM + L + X + X + X + IX
= 1000 + 900 + 50 + 10 + 10 + 10 + 9 = 1989

MMMDCLXXVIII
= M + M + M + D + C + L + X + X + V + III
= 1000 + 1000 + 1000 + 500 + 100 + 50 + 10 + 10 + 5 + 3 = 3678

4.1. **Express in the Roman numbers theses numbers: 94, 952, 1697, and 2069.**

94 = 90 + 4 = XC + IV = XCIV
952 = 900 + 50 + 2 = CM + L + II = CMLII
1697 = 1000 + 900 + 50 + 40 + 7 = M + CM + L + XL + VII = MCMLXLVII
2069 = 2000 + 40 + 9 = 1000 + 1000 + 50 + 10 + 9 = M + M + L + X + IX = MMLXIX.

Examples N

0.   What is the thousand's digit in a number made of 97 of tens, 98 of hundreds, 95 of ones, and 39 of thousands?

1.   Answering the questions below, use the numbers as follows:

9, 5, 2, 0, 4, 7, and 6

1.0.   What is the 24th largest integer if we use each number once only?

1.1.   What is the smallest 7-digit integer if we use each number once only?

1.2.   What is the 18th smallest 7-digit integer if we use each number once only?

1.3.   What is the smallest integer if we use all the numbers but once only?

1.4.   What is the 4th largest number if we can use each number up to twice?

2.   Find all possible integers that can be put in the blank below:

6900_080 > 69005081

3.   What is the number that is the sum of 25 of 10's, 37 of 1's, 97 of 0.1's, 85 of 0.01's, 69 of 0.001's, and 78 of 0.0001's?

4.   Make the second biggest integer and the second smallest integer using twice each all the numbers as follows: 0, 3, and 2.

6.   Using the 8 numbers below once each, make the integer closest to 75000000.

4, 7, 0, 1, 5, 3, 8, and 2

7.   Four people are making numbers using two sets of cards each. Each set is made of 10 different cards, each of which has one of 0, 1, 2, . . . 8 and 9. And each person has put 5 of the cards the way below:

Jane: 7, _, 4, 2, 0, 3, _   Mike: 7, 8, 9, _, 4, 6, _
Robert: 7, _, _, 6, 5, 1, 8   Becky: 7, 9, _, 8, _, 4, 3

Who then, can make the largest number filling in the blanks?

Suggestions or Solutions
To the Problems in the Examples N

0.   What is the thousand's digit in a number made of 97 of tens, 98 of hundreds, 95 of ones, and 39 of thousands?

Getting the sum of the numbers, we get:

$39 \times 1000 + 98 \times 100 + 97 \times 10 + 95 \times 1 = 39000 + 9800 + 970 + 95$

$= 40000 - 1000 + 10000 - 200 + 1000 - 30 + 100 - 5$

$= 50000 - 135 = 49800 + 65 = 49865$. So the 1000's digit is 9.

1.   Answering the questions below, use the numbers as follows:
9, 5, 2, 0, 4, 7, and 6

1.0.   What is the 80th largest integer if we use each number once only?

The largest integer will be a 7-digit integer.
And putting such integers in descending order, we can put them the way below:

9765420, 9765402,
9765240, 9765204, 9765042, 9765024,
9764520, 9764502, 9764250, 9764205, 9764052, 9764025,
9762540, 9762504, 9762450, 9762405, 9762054, 9762045,
9760542, 9760524, 9760452, 9760425, 9760254, 9760245,
9756420, 9756402, 9756240, 9756204, 9756042, 9756024, 9754620, 9754602,
9754260, 9754206, 9754062, 9754026, 9752640, 9752604, 9752460, 9752406,
9752064, 9752046, 9750642, 9750624, 9750462, 9750426, 9750264, 9750246,
9746520, 9746502, 9746250, 9746205, 9746052, 9746025, 9745620, 9745602,
9745260, 9745206, 9745062, 9745026, 9742650, 9742605, 9742560, 9742506,
9742065, 9742056, 9740652, 9740625, 9740562, 9740526, 9740265, 9740256,
9726540, 9726504, 9726450, 9726405, 9726054, 9726045, 9725640, <u>9725604</u>.

1.1.   What is the smallest 7-digit integer if we use each number once only?

We are given 7 single digit numbers, and of those, the smallest nonzero is 2.
So the smallest 7-digit integer begins with 2, that is, the highest digit is 2.
And the smaller the digit is, the smaller the number if the number is positive, of course.
Thus, the smallest 7-digit integer is 2045679.

1.2.   What is the 18th smallest 7-digit integer if we use each number once only?

The largest integer will be a 7-digit integer.
And putting such integers in ascending order, we can put them the way below:

2045679, 2045697, 2045769, 2045796, 2045967, 2045976,
2046579, 2046597, 2046759, 2046795, 2046957, 2046975,
2047569, 2047596, 2047659, 2047695, 2047956, **2047965**.

1.3.   What is the smallest integer if we use all the numbers but once only?

The same numbers can look different. So for instance, we can have: 0001502 = 1502,
and $\frac{4}{2}$ = 2. And thus, the smallest integer is 0245679, which is 245679, of course.

1.4.   What is the 4th largest number if we can use each number up to twice?

The largest number is 99776655442200.
The second largest number is 99776655442020.
The third largest number is 99776655442002.
And the fourth largest number is 99776655440220.

2.   Find all possible integers that can be put in the blank below:
6900_080 > 69005081

The number in the 1000's digit has to be greater than 5.
Therefore, we can put in the blank any of 6, 7, 8, and 9

3.   What is the number that is the sum of 25 of 10's, 37 of 1's, 97 of 0.1's, 85 of 0.01's, 69 of 0.001's, and 78 of 0.0001's?

10 x 25 + 1 x 37 + 0.1 x 97 + 0.01 x 85 + 0.001 x 69 + 0.0001 x 78
= 250 + 37 + 9.7 + 0.85 + 0.069 + 0.0078
= 280 + 7 + 9 + 0.7 + 0.8 + 0.05 + 0.06 + 0.009 + 0.007 + 0.0008
= 280 + 16 + 1.5 + 0.11 + 0.016 + 0.0008
= 297 + 0.5 + 0.1 + 0.01 + 0.01 + 0.006 + 0.0008
= 297 + 0.6 + 0.02 + 0.006 + 0.0008
= 297.6268.

4.   Make the second biggest integer and the second smallest integer using twice each all the numbers as follows: 0, 3, and 2.

The biggest is 332200.
And the second biggest is 332020.

The smallest is 002233, which is 2233, of course.
And the second smallest is 002323, which is of course, 2323.

6.   Using the 8 numbers below once each, make the integer closest to 75000000.

4, 7, 0, 1, 5, 3, 8, and 2

We have 7, 5 and 0 in 75000000. And we have 7, 5 and 0 in the list of the numbers. So using 7 as the highest digit, using 5 as the second highest digit, and using 0 as the third highest, we get *750*00000. Then, we are left with5 numbers: 4, 1, 3, 8, and 2. What number then, do we need to make?

It is the smallest 5-digit integer can be made of the 5 numbers left. And the smallest will be 12348. So the integer we want is: 75000000 + 12348 = 75012348.

7. **Four people are making numbers using two sets of cards each. Each set is made of 10 different cards, each of which has one of 0, 1, 2, . . . 8 and 9. And each person has put 5 of the cards the way below:**

Jane: 7, _, 4, 2, 0, 3, _   **Mike: 7, 8, 9, _, 4, 6, _**

Robert: 7, _, _, 6, 5, 1, 8   **Becky: 7, 9, _, 8, _, 4, 3**

Who then, can make the largest number filling in the blanks?

Each person has used 5 cards. So they now have 15 cards each.

Jane has 2 of 1s, 5s, 6s, 8s, and 9s each, and 1 of 0s, 2s, 3s, and 4s, and 7s each.
That is, she has now: 1, 1, 5, 5, 6, 6, 8, 8, 9, 9, 0, 2, 3, 4, and 7.
So the biggest number she can make is 7942039.

Mike has 2 of 0s, 1s, 2s, 3s, and 5s each, and 1 of 4s, 6s, 7s, 8s, and 9s each.
So the biggest number he can make is 7899468.

Robert has 2 of 0s, 2s, 3s, 4s, and 9s each, and 1 of 1s, 5s, 6s, 7s, and 8s each.
So the biggest number he can make is 7996518.

Becky has 2 of 0s, 1s, 2s, 5s, and 6s each, and 1 of 3s, 4s, 7s, 8s, and 9s each.
So the biggest number she can make is 7998843.

And thus, Becky should be able to make the largest number.

Examples O

Examples O

0.   Rearrange in ascending (increasing) order, the 4-digit numbers below:

7?52, 7?94, 79?6, 779?, 7?01, and 78?0, where ? = 0, 1, 2 . . . 8, or 9.

Substitute the question mark with the number assigned to the mark.
For instance, if **? = 0**, then:
7?52, 7?94, 79?6, 779?, 7?01, and 78?0 will be: 7052, 7094, 7906, 7790, 7001, and 7800.
So next, rearranging them in ascending order, we get: 7906, 7800, ...

0.A.   ? = 0

0.B.   ? = 1

0.C.   ? = 2

0.D.   ? = 3

0.E.   ? = 4

0.F.   ? = 5

0.G.   ? = 6

0.H.   ? = 7

0.I.   ? = 8

0.J.   ? = 9

1.   Wily has his ID, which is a four-digit number. The number is less than 7000, and has 3, 5, 7, and 9 in each digit. The digits 5 and 9 are consecutive (next to each other), 3 and 5 are apart, and 3 and 7 are apart, too. What then, is his ID number?

2.   Suppose a company is getting paid $7_40_5 a month, and out of the monthly income, the company is spending $532758 a month. Then:

2.A.   How much at most, can the company save per month?

2.B.   How much at least, can the company save per month?

3.   We are given two sets of cards. And each set has 10 different cards, each of which has one of 0, 1, 2 . . . 8, and 9.

3.0.   Find the biggest number we can make using 9 cards out of the 20 cards.

3.1.   Suppose 9 cards have been removed from the 20 cards. What then, can be the smallest number we can make using 8 cards from the 11cards left?

4.   Two people made numbers using number cards. Each was given two sets of cards. Each set has 10 different cards, each of which has one of 0, 1, 2 . . . 8, and 9. One person made a number 44 million less than the biggest number that can be made of 9 cards from the 20 cards. And the number made by the other person is 406 million and 400 bigger than the smallest number that can be made of 8 cards out of the 20 cards. What then, is the difference between the two numbers made?

Suggestions or Solutions
To the Problems in the Examples O

0. **Rearrange in ascending order, the 4-digit numbers below:**

7?52, 7?94, 79?6, 779?, 7?01, and 78?0, where ? = 0, 1, 2 . . . 8, or 9.

0.A. ? = 0

7052, 7094, 7906, 7790, 7001, 7800. So we get: 7906, 7800, 7790, 7094, 7052, 7001

0.B. ? = 1

7152, 7194, 7916, 7791, 7101, 7810. So we get: 7916, 7810, 7791, 7194, 7152, 7101

0.C. ? = 2

7252, 7294, 7926, 7792, 7201, 7820. So we get: 7926, 7820, 7792, 7294, 7252, 7201

0.D. ? = 3

7352, 7394, 7936, 7793, 7301, 7830. So we get: 7936, 7830, 7793, 7394, 7352, 7301

0.E. ? = 4

7452, 7494, 7946, 7794, 7401, 7840. So we get: 7946, 7840, 7794, 7494, 7452, 7401

0.F. ? = 5

7552, 7594, 7956, 7795, 7501, 7850. So we get: 7956, 7850, 7795, 7594, 7552, 7501

0.G. ? = 6

7652, 7694, 7966, 7796, 7601, 7860. So we get: 7966, 7860, 7796, 7694, 7652, 7601

0.H. ? = 7

7752, 7794, 7976, 7797, 7701, 7870. So we get: 7976, 7870, 7797, 7794, 7752, 7701

0.I. ? = 8

7852, 7894, 7986, 7798, 7801, 7880. So we get: 7986, 7894, 7880, 7852, 7801, 7798

0.J. ? = 9

7952, 7994, 7996, 7799, 7901, 7890. So we get: 7996, 7994, 7952, 7901, 7890, 7799

1. **Wily has his ID, which is a four-digit number. The number is less than 7000, and has 3, 5, 7, and 9 in each digit. The digits 5 and 9 are consecutive (next to each other), 3 and 5 are apart, and 3 and 7 are apart, too. What then, is his ID number?**

To begin with, we know: ID < 7000. So the highest digit ≤ 6.
Next, each digit is one of 3, 5, 7 and 9. So the highest digit is 3 or 5.

Suppose first, the highest digit is 3. Then, the rest of the digits are: 5, 7, and 9.
And we know: 3 and 5 are apart, and 3 and 7 are apart, too.
So the ID can be: 3957 or 3975.
And also, we know 5 and 9 are consecutive. So 5 and 9 have to be put this way: 59 or 95.
And thus, if the highest digit is 3, the ID has to be 3957.

Suppose next, the highest digit is 5. Then, the rest of the digits are: 3, 7, and 9.
And we know: 3 and 5 are apart, and 3 and 7 are apart, too.
So the ID can be 5793 only.
We know however, 5 and 9 have to be put this way only: 59 or 95.
So if the highest digit cannot be 5.

And thus, the ID must be 3957.

2. **Suppose a company is getting paid $7_40_5 a month, and out of the monthly income, the company is spending $532758 a month. Then:**

2.A. **How much at most, can the company save per month?**

The maximum that the company can get paid is $794095. So the amount the company can save at most each month is: $794095 − $532758 = $261337.

2.B. **How much at least, can the company save per month?**

The minimum that the company can make is $704005. So the amount the company can save at least each month is: $704005 − $532758 = $171247

3. We are given two sets of cards. And each set has 10 different cards, each of which has one of 0, 1, 2 . . . 8, and 9.

3.0. Find the biggest number we can make using 9 cards out of the 20 cards.

The biggest is 998877665.

3.1. Suppose 9 cards have been removed from the 20 cards. What then, can be the smallest number we can make using 8 cards from the 11cards left?

We don't know what 9 cards have been removed. We can see however, what 8 cards have to be used if we have to make the smallest of all the numbers that can be made using 8 cards. So it doesn't matter that we remove 9 cards or not.
What matters is that we use 8 cards among the 20 cards.
So the smallest number that can be made is 00112233, which is 112233.

4. Two people made numbers using number cards. Each was given two sets of cards. Each set has 10 different cards, each of which has one of 0, 1, 2 . . . 8, and 9. One person made a number 44 million less than the biggest number that can be made of 9 cards from the 20 cards. And the number made by the other person is 406 million and 400 bigger than the smallest number that can be made of 8 cards out of the 20 cards. What then, is the difference between the two numbers made?

The biggest number that can be made of 9 cards is 998877665.
So the number made by one of the two people is: 998877665 − 44000000 = 954877665.

Next, the smallest number that can be made of 8 cards is 00112233, which is 112233.
So the number made by the other person is: 112233 + 406000400 = 406112633.

And thus, the difference is: 954877665 − 406112633 = 548765032.

Examples P

0.   If we use each of 2 and 5 once only to make a 2-digit integer, how many different 2-digit integers can we make?

1.   If we use each of 4, 3, and 7 once only to make a 3-digit integer, how many different 3-digit integers can we make?

2.   If we use each of 8, 0, and 6 once only to make a 3-digit integer, how many different 3-digit integers can we make?

3.   If we use each of 9, 3, 1, and 7 once only to make a 4-digit integer, how many different 4-digit integers can we make?

4.   If we use each of 6, 8, 0, and 4 once only to make a 4-digit integer, how many different 4-digit integers can we make?

5.   If we use each of 2, 5, and 7 once only to make a 2-digit integer, how many different 2-digit integers can we make?

Suggestions or Solutions
To the Problems in the Examples P

0. If we use each of 2 and 5 once only to make a 2-digit integer, how many different 2-digit integers can we make?

In a 2-digit integer, the highest digit is 10's digit.

And each of the numbers given has to be used only once.

So we can make 25 and 52, and thus, can make 2 different 2-digit integers.

1. If we use each of 4, 3, and 7 once only to make a 3-digit integer, how many different 3-digit integers can we make?

In a 3-digit integer, the highest digit is 100's digit.
And each number given has to be used only once.

So we have 3 choices for the 100's digit.

Then, we get 2 choices for the 10's digit.

And then, we get 1 choice for the 1's digit.

So we get: 3 x 2 x 1 = 6 choices, and thus, can make 6 different 3-digit integers.

2.   If we use each of 8, 0, and 6 once only to make a 3-digit integer, how many different 3-digit integers can we make?

In a 3-digit integer, the highest digit is 100's digit.
And each number given has to be used only once.
And we have 0 in the list of the numbers to be used.

We cannot use though, 0 for the 100's digit. So we have 2 choices for the 100's digit.

We can use however, 0 for any of the other lower digits. So we get 2 choices again, for the10's digit. Then, we get 1 choice for the 1's digit.

So we get: 2 x 2 x 1 = 4 choices, and thus, can make 4 different 3-digit integers.

3.   If we use each of 9, 3, 1, and 7 once only to make a 4-digit integer, how many different 4-digit integers can we make?

In a 4-digit integer, the highest digit is 1000's digit.
And each number given has to be used once only.

So first, we have 4 choices for the 1000's digit.

Then, we get 3 choices for the 100's digit.

Then, we get 2 choices for the 10's digit.

And then, we get 1 choice for the 1's digit.

So we get: 4 x 3 x 2 x 1 = 24 choices, and thus, can make 24 different 4-digit integers.

4. If we use each of 6, 8, 0, and 4 once only to make a 4-digit integer, how many different 4-digit integers can we make?

In a 4-digit integer, the highest digit is 1000's digit.
Each number has to be used only once.
And we have 0 in the list of the numbers to be used, but cannot use it for the 1000's digit.

So we have 3 choices for the 1000's digit.

We can use though, 0 for any of the other lower digits.
So we get 3 choices again, for the 100's digit.

Then, we get 2 choice for 10's digit.

And then, we get 1 choice for the 1's digit.

So we get: 3 x 3 x 2 x 1 = 18 choices, and thus, can make 18 different 4-digit integers.

5. If we use each of 2, 5, and 7 once only to make a 2-digit integer, how many different 2-digit integers can we make?

In a 2-digit integer, the highest digit is 10's digit.
We have three different numbers to choose from.
And each number has to be used once only.

So we have 3 choices for the 10's digit.

Then, we get 2 choices for the 1's digit.

So we get: 3 x 2 = 6 choices, and thus, can make 6 different 2-digit integers.

Examples Q

0. Using each of 3, 0, and 8 once only, how many 2-digit integers can we make?

1. Using each of 5, 1, 4, and 7 once only, how many 3-digit integers can we make?

2. Using each of 4, 6, 0, and 9 once only, how many 3-digit integers can we make?

3. Using once only each of 9 numbers 1, 2, ... 8, and 9, how many 9-digit integers can we make?

4. Using once only each of 10 numbers 0, 1, 2, ... 8, and 9, how many 10-digit integers can we make?

5. Using once only each of 9 numbers 1, 2, ... 8, and 9, how many 6-digit integers can we make?

6. Using once only each of 10 numbers 0, 1, 2, ... 8, and 9, how many 8-digit integers can we make?

7. Using once only each of 3 numbers 1, 1, and 2, how many 3-digit integers can we make?

8. Using once only each of 4 numbers 1, 1, 4, and 4, how many 4-digit integers can we make?

9. Using once only each of 5 numbers 1, 1, 4, 4, and 4, how many 5-digit integers can we make?

A. Using once only each of 7 numbers 1, 1, 3, 3, 4, 4, and 4, how many 7-digit integers can we make?

Suggestions or Solutions
To the Problems in the Examples Q

0.   Using each of 3, 0, and 8 once only, how many 2-digit integers can we make?

   We can make 4 different 2-digit integers.

If not sure of how to get it, follow the steps below:

In a 2-digit integer, the highest digit is 10's digit.
And each number given has to be used only once.
And we have 0 in the list of the numbers to be used, but cannot use it for the 10's digit.

So we have 2 choices for the 10's digit.

We can use however, 0 for the other digit.

So we get 2 choices again, for the 1's digit.

So we get: 2 x 2 = 4 choices, and thus, can make 4 different 2-digit integers.

1.   Using each of 5, 1, 4, and 7 once only, how many 3-digit integers can we make?

   We can make 24 different 3-digit integers.

If not sure of how to get it, follow the steps below:

In a 3-digit integer, the highest digit is 100's digit.
And each number given has to be used once only.

So first, we have 4 choices for the 100's digit.

Then, we get 3 choices for the 10's digit.

Then, we get 2 choices for the 1's digit.

So we get: 4 x 3 x 2 = 24 choices, and thus, can make 24 different 3-digit integers.

2.   Using each of 4, 6, 0, and 9 once only, how many 3-digit integers can we make?

We can make 18 different 3-digit integers.

If not sure of how to get it, follow the steps below:

In a 3-digit integer, the highest digit is 100's digit.
Each number has to be used only once.
And we have 0 in the list of the numbers to be used, but cannot use it for the 100's digit.

So we have 3 choices for the 100's digit.

We can use though, 0 for any of the other lower digits.

So we get 3 choices again, for the 10's digit.

Then, we get 2 choice for 1's digit.

So we get: 3 x 3 x 2 = 18 choices, and thus, can make 18 different 3-digit integers.

3.   Using once only each of 9 numbers 1, 2, … 8, and 9, how many 9-digit integers can we make?

9!, that is, 362880 different 9-digit integers.

If not sure of how to get it, follow the steps below:

In a 9-digit integer, the highest digit is 10^8's digit, i.e., 100000000's digit.
And each number given has to be used once only.

So first, we have 9 choices for the 10^8's digit.
Then, we get 8 choices for the 10^7's digit.
Then, we get 7 choices for the 10^6's digit.
And so forth. So we get: 9 x 8 x 7 x … x 2 x 1 choices.
And 9 x 8 x 7 x … x 2 x 1 is called 9! read as 9 factorial. And 9! = 362880.

And thus, we can make 9! different 9-digit integers, that is, 362880 9-digit integers.

4. Using once only each of 10 numbers 0, 1, 2, … 8, and 9, how many 10-digit integers can we make?

  9 x 9! different 10-digit integers.   *If not sure of how to get it, follow the steps below:*

In a 10-digit integer, the highest digit is 10^9's digit, i.e., 1000000000's digit.
And each number given has to be used once only.
We have 0 in the list of the numbers to be used, but cannot use it for the 10^9's digit.

So first, we have 9 choices for the 10^9's digit.

We can use though, 0 for any of the other lower digits.
So we get 9 choices again, for the 10^8's digit.

Then, we get 8 choices for the 10^7's digit.

Then, we get 7 choices for the 10^6's digit.
So we get: 9 x 9 x 8 x 7 x … x 2 x 1 = 9 x 9! choices.

And thus, we can make 9 x 9! different 10-digit integers.

5. Using once only each of 9 numbers 1, 2, … 8, and 9, how many 6-digit integers can we make?

  9!/3!, that is, 60480 different 6-digit integers.

If not sure of how to get it, follow the steps below:

In a 6-digit integer, the highest digit is 10^5's digit, i.e., 100000's digit.
And each number given has to be used once only.
So first, we have 9 choices for the 10^5's digit.
Then, we get 8 choices for the 10^4's digit.
Then, we get 7 choices for the 10^3's digit. And so forth.
So we get: 9 x 8 x 7 x 6 x 5 x 4 = 9!/3! choices. And 9!/3! = 362880/6 = 60480.

And thus, we can make 9!/3!, that is, 60480 different 6-digit integers.

6. **Using once only each of 10 numbers 0, 1, 2, … 8, and 9, how many 8-digit integers can we make?**

9 x 9!/2, that is, 1632960 different 8-digit integers.

If not sure of how to get it, follow the steps below:

In an 8-digit integer, the highest digit is 10^7's digit, i.e., 10000000's digit.
And each number given has to be used once only.
We have 0 in the list of the numbers to be used, but cannot use it for the 10^7's digit.

So first, we have 9 choices for the 10^7's digit.

We can use though, 0 for any of the other lower digits.
So we get 9 choices again, for the 10^6's digit.

Then, we get 8 choices for the 10^5's digit.

Then, we get 7 choices for the 10^4's digit.

So we get: 9 x 9 x 8 x 7 x 6 x 5 x 4 x 3 = 9 x 9!/2! = 9 x 9!/2 = 9 x 362880/2

= 9 x 181440 = 1632960 choices.

And thus, we can make 9 x 9!/2 different 8-digit integers, that is, 1632960 different 8-digit integers.

7. Using once only each of 3 numbers 1, 1, and 2, how many 3-digit integers can we make?

We can make 3 different 3-digit integers.

If not sure of how to get it, follow the steps below:

Assuming for instance, we use once only each of three different numbers 1, 2, and 3 to make a 3-digit integer, we get: 3 x 2 x 1 = 3! = 6 choices, and thus, can make 6 different 3-digit integers. And all the six integers are as follows: 123, 132, 213, 231, 312, 321.

Now, replacing all the 3s above with 1s, we get: 121, 112, 211, 211, 112, 121.

Then, we can see that there are only three different 3-digit integers. That is to say that each different integer appears twice.

So using once only each of 3 numbers 1, 1, and 2, we can make 3 different 3-digit integers, and the integers are: 112, 121, 211.

And thus, assuming we get x different 3-digit integers if we use once only each of 3 numbers, two of which are the same, we get: x x 2! = y, where y is the number of all the 3-digit integers we can make using once only each of 3 different nonzero numbers.

And we know $y = 3!$.
So we get: x x 2! = 3!.

Thus, we get: $x = 3!/2!$, which is 3.

8.   Using once only each of 4 numbers 1, 1, 4, and 4, how many 4-digit integers can we make?

We can make 6 different 4-digit integers.

If not sure of how to get it, follow the steps below:

Assuming for instance, we use once only each of 4 different numbers 1, 2, 3, and 4 to make a 4-digit integer, we get: 4 x 3 x 2 x 1 = 4! = 24 choices, and thus, can make 24 different 4-digit integers. And all the 24 integers are as follows:

1234, 1243, 1324, 1342, 1423, 1432.   2134, 2143, 2314, 2341, 2413, 2431.
3124, 3142, 3214, 3241, 3412, 3421.   4123, 4132, 4213, 4231, 4312, 4321.

Now, replacing all the 2s above with 1s, we get:

1134, 1143, 1314, 1341, 1413, 1431.   1134, 1143, 1314, 1341, 1413, 1431.
3114, 3141, 3114, 3141, **3411,** 3411.   **4113, 4131,** 4113, 4131, **4311,** 4311.

Then, we can see that there are only 12 different 4-digit integers. That is to say that each different integer appears twice.
So using once only each of 4 numbers 1, 1, 3, and 4, we can make 12 different 4-digit integers, and the integers are as follows:

1134, 1143, 1314, 1341, 1413, 1431.
3114, 3141, 3411, 4113, 4131, 4311.

Now, replacing all the 3s above with 4s, we get:

1144, 1144, **1414, 1441,** 1414, 1441.
4114, 4141, 4411, 4114, 4141, 4411.

Then, we can see that there are only 6 different 4-digit integers. That is to say that each different integer appears twice.

So using once only each of 4 numbers 1, 1, 4 and 4, we can make 6 different 4-digit integers, and the integers are as follows: 1144, 1414, 1441, 4411, 4141, 4114,

So what pattern can you see?

Suppose we get x different 4-digit integers if we use once only each of 4 nonzero numbers, and two of the four are the same as 1, 1, 3, and 4.

Then, we get: $x \times 2! = z$, where z is the number of all 4-digit integers we can make using once only each of 4 different nonzero numbers.

And we know $z = 4!$. So we get: $x \times 2! = 4!$. Thus, we get: $x = 4!/2!$, which is 12.

Suppose this time, we get y different 4-digit integers if we use once only each of 4 nonzero numbers, two of which are the same as each other, and the other two are the same as each other, too.

Then, we get: $y \times 2! = x$, where x is the number of all 4-digit integers we can make using once only each of 4 nonzero numbers, two of which are the same.

And we know $x = 4!/2!$. So we get: $y \times 2! = 4!/2!$.

Thus, we get: $y = (4!/2!)/2! = 4!/2!/2! = 4!/(2! \times 2!)$, which is 6.

9. Using once only each of 5 numbers 1, 1, 4, 4, and 4, how many 5-digit integers can we make?

We can make 10 different 5-digit integers.

If not sure of how to get it, follow the steps below:

Suppose we get x different 5-digit integers if we use once only each of 5 nonzero numbers, two of which are the same. Then, we get: $x \times 2! = z$, where z is the number of all 5-digit integers we can make using once only each of 5 different nonzero numbers.

And we know $z = 5!$. So we get: $x \times 2! = 5!$. Thus, we get: $x = 5!/2!$, which is 60.

Suppose next, we get y different 5-digit integers if we use once only each of 5 nonzero numbers, two of which are the same as each other, and the other three are the same as each other, too. Then, we get: $y \times 3! = x$, where x is the number of all 5-digit integers we can make using once only each of 5 nonzero numbers, two of which are the same.

And we know $x = 5!/2!$. So we get: $y \times 3! = 5!/2!$.

Thus, we get: $y = (5!/2!)/3! = 5!/2!/3! = 5!/(2! \times 3!)$, which is 10.

A.   Using once only each of 7 numbers 1, 1, 3, 3, 4, 4, and 4, how many 7-digit integers can we make?

We can make $7!/(2! \times 2! \times 3!)$, that is, 210 different 7-digit integers.

If not sure of how to get it, follow the steps below:

Suppose we get x different 7-digit integers if we use once only each of 7 nonzero numbers, two of which are the same.

Then, we get: $x \times 2! = z$, where z is the number of all 7-digit integers we can make using once only each of 7 different nonzero numbers.

And we know $z = 7!$. So we get: $x \times 2! = 7!$. Thus, we get: $x = 7!/2!$, which is 2520.

Suppose next, we get *y* different 7-digit integers if we use once only each of 7 nonzero numbers, two of which are the same as each other, and two of the other five are the same as each other, too.

Then, we get: $y \times 2! = x$, where *x* is the number of all 7-digit integers we can make using once only each of 7 nonzero numbers, two of which are the same.

And we know $x = 7!/2!$. So we get: $y \times 2! = 7!/2!$.

Thus, we get: $y = (7!/2!)/2! = 7!/2!/2! = 7!/(2! \times 2!)$, which is 1260.

Suppose this time, we get *w* different 7-digit integers if we use once only each of 7 nonzero numbers, two of which are the same as each other, two of the other five are the same as each other, and the other three are the same as each other, too.

Then, we get: $w \times 3! = y$, where *y* is the number of all 7-digit integers we can make using once only each of 7 nonzero numbers, two of which are the same, and two of the other five are the same as each other, too.

And we know $y = 7!/(2! \times 2!)$. So we get: $w \times 3! = 7!/(2! \times 2!)$.

Thus, we get: $w = \{7!/(2! \times 2!)\}/3! = 7!/(2! \times 2! \times 3!)$, which is 210.

What if we use *n* numbers, *u* of which are the same, *v* of which are the same, *w* of which are the same, *x* of which are the same, and *y* of which are the same?

If we make *n*-digit integers, we can make $n!/(u! \times v! \times w! \times x! \times y!)$ different *n*-digit integers.

Examples R

0. Find all the integers that satisfy all the statements below.

   A. They are 2-digit integers positive.
   B. 4 and 6 can divide the integers.
   C. In each of the integers, the one's digit is twice the ten's digit.

1. Find all the integers that satisfy all below.

   A. They are 3-digit integers positive.
   B. 3, 4, 5 and 6 can divide the integers.
   C. In each of the integers, the ten's digit is twice the hundred's digit.

2. Find the integer described as follows.

   A. It is a 2-digit integer positive.
   B. Adding together the two numbers in both digits, we get an 8.
   C. It is an odd integer, and is closest to 40.

3. Find the date of birth described as follows.

   A. The date is a 2-digit odd integer, and the difference between the 2 digits is 7.
   B. The 100's digit of the year is the same as one of the two digits in the date.
   C. The 1's digit in the month is 0.
   D. In the year, the sum of the numbers in the 1000's and the 100's digits is equal to the sum of the numbers in the other digits.
   E. And in the year, the 1's digit is less than the 10's digit, and the year is made of odd integers.

Suggestions or Solutions
To the Problems in the Examples R

0. **Find all the integers that satisfy all the statements below.**

    A. They are 2-digit integers positive.

    B. 4 and 6 can divide the integers.

    C. In each of the integers, the one's digit is twice the ten's digit.

12, 24, 36, or 48	*If not sure of how to get it, follow the steps below:*

• From the statement **A**, we can say that:

Assuming N is such an integer, we can set: $N = 10a + b$, where $1 \le a \le 9$, $0 \le b \le 9$, and a and b are integers. Why not $0 \le a \le 9$ but $1 \le a \le 9$ though?

That's because N is a 2-digit integer positive. If $a = 0$, N cannot be a 2-digit integer.

• From the statement **B**, we can say that:

N is a multiple of a number. What then, is the number?

We have: $4 = 2 \times 2 = 2^2$, and $6 = 2 \times 3$.

So the number is: $2^2 \times 3 = 12$, which is the least common multiple of 4 and 6.

And the least common multiple is often called the **LCM**, which is the acronym.

And thus, we can set: $N = 12n$ where n is an integer ≥ 1.

• From the statement **C**, we can say that:

We know a is the 10's digit, and b is the 1's digit. So we get: $b = 2a$.

So putting threads together:

To begin with, in the statement **B**, we have: $N = 12n$ for $n = 1, 2, 3, 4 \ldots$

From **A** though, we get: $\mathbf{10 \leq N \leq 99}$.

So we get: $N = 12, 24, 36, 48, 60, 72, 84, 96$.

And in the **C**, we have: $b = 2a$ where a is the 10's digit, and b is the 1's digit.

And thus, we get: $N = 12, 24, 36$, or 48.

1. **Find all the integers that satisfy all below.**

   A. **They are 3-digit integers positive.**
   B. **3, 4, 5 and 6 can divide the integers.**
   C. **In each of the integers, the ten's digit is twice the hundred's digit.**

   120, 240, 360, or 480   *If not sure of how to get it, follow the steps below:*

● From the statement **A**, we can say that:

Assuming N is such an integer, we can set: $N = 100a + 10b + c$, where a, b, and c are integers, and $\mathbf{1 \leq a \leq 9, 0 \leq b \leq 9}$, and $\mathbf{0 \leq c \leq 9}$. Why $\mathbf{1 \leq a \leq 9}$, though?

That's because N is a 3-digit integer positive.

● From the statement **B**, we can say that:

N is a multiple of a number. What then, is the number?

It is the least common multiple of all the divisors: 3, 4, 5 and 6.

That is, it is the **LCM** of 3, 4, 5 and 6.   What then, is the **LCM**?

We have: $4 = 2 \times 2 = 2^2$, and $6 = 2 \times 3$. So the **LCM** is: $2^2 \times 3 \times 5 = 60$.

And thus, we can set: $N = 60n$ where n is an integer ≥ 2.

Why not: n is an integer ≥ 1 but: n is an integer ≥ 2, though?

It's because N is a 3-digit integer.

• From the statement **C**, we can say that:

We know a is the 100's digit, and b is the 10's digit. So we get: $b = 2a$.

So putting threads together:

To begin with, in the statement **B**, we have: $N = 60n$ for $n = 2, 3, 4 \ldots$

From **A** though, we get: $100 \leq N \leq 999$.
So we get: $N = 120, 180, 240, 300, 360, \ldots 900, 960$.

And in the **C**, we have: $b = 2a$ where a is the 100's digit, and b is the 10's digit.

And thus, we get: $N = 120, 240, 360,$ or 480.

2. Find the integer described as follows.

    A. It is a 2-digit integer positive.

    B. Adding together the two numbers in both digits, we get an 8.

    C. It is an odd integer, and is closest to 40.

It is 35.   *If not sure of how to get it, follow the steps below:*

• From the statement **A**, we can say that:

Assuming N is the integer, we can set: $N = 10a + b$, where $1 \leq a \leq 9$, $0 \leq b \leq 9$, and a and b are integers.

• From the statement **B**, we can say that:

The two numbers in both digits are a and b.   So we get: $a + b = 8$.

And we have: $1 \leq a \leq 9$, and $0 \leq b \leq 9$, where a and b are integers.

So we can set: $(a, b) = (1, 7), (2, 6), (3, 5), (4, 4), (5, 3), (6, 2), (7, 1)$.

• And the statement **C** says: N is an odd integer.

And in **A**, we have: $N = 10a + b$. So b is an odd integer.

And we have: $0 \leq b \leq 9$.   So we can set: $b = 1, 3, 5, 7, 9$.

And in **B**, we have: $(a, b) = (1, 7), (2, 6), (3, 5), (4, 4), (5, 3), (6, 2), (7, 1)$.

So we can reduce it this way: $(a, b) = (1, 7), (3, 5), (5, 3), (7, 1)$.

And thus, N is one of the integers as follows: 17, 35, 53, 71.

• And also, the statement **C** says: N is closest to 40.

So we can see that $N = 35$.

3.   **Find the date of birth described as follows.**

    A. **The date is a 2-digit odd integer, and the difference between the 2 digits is 7.**
    B. **The 100's digit of the year is the same as one of the two digits in the date.**
    C. **The 1's digit in the month is 0.**
    D. **In the year, the sum of the numbers in the 1000's and the 100's digits is equal to the sum of the numbers in the other digits.**
    E. **And in the year, the 1's digit is less than the 10's digit, and the year is made of odd integers.**

Oct 29, 1991 or 1973.   *If not sure of how to get it, follow the steps below:*

To begin with, a date of birth is composed of the date, the month and the year.

    A. The date is a 2-digit odd integer, and the difference between the 2 digits is 7.

First, assuming the date is d, we can set: $1 \le d \le 31$.
Next, the date is a 2-digit odd integer. So we can set: $d = 11, 13, 15, \dots$ or 31.
And also, we can set it this way: $d = 2k + 11$, where $k = 0, 1, 2, \dots 10$.
And we can put it this way, too: $d = 2k + 1$, where $k = 5, 6, 7, \dots 15$.

Next, the difference between the 2 digits is 7. So we get: $d = 29$.
And therefore, the date must be 29.

    B. The 100's digit of the year is the same as one of the two digits in the date.

So assuming the year is $wxyz$, where w, x, y, and z are digits, we get: $x = 2$ or 9.

    C. The 1's digit in the month is 0.

First, assuming the month is m, we can set: $1 \le m \le 12$.
And the 1's digit in the month is 0. So we get: $m = 10$, which is the only choice.
And therefore, the month must be October.

D. In the year, the sum of the numbers in the 1000's and the 100's digits is equal to the sum of the numbers in the other digits.

In B, we assume that the year is $wxyz$, where w, x, y, and z are digits.
So w is the 1000's digit, and x is the 100's digit.
And thus, we get: $w + x = y + z$. And in B, we have: $x = 2$ or 9.
So we get: $w + 2 = y + z$, or $w + 9 = y + z$.

E. And in the year, the 1's digit is less than the 10's digit, and the year is made of odd integers.

Now, the year is made of odd integers.
So of the two cases where $w + 2 = y + z$, and $w + 9 = y + z$, we want to use: $w + 9 = y + z$.

Next, in the year, the 1's digit is less than the 10's digit.
We know y is the 10's digit, z is the 1's digit. So we get: $y > z$.

And we know the year is made of odd integers. So we get:

$y = 9 \Rightarrow z = 1, 3, 5, 7$
$y = 7 \Rightarrow z = 1, 3, 5$
$y = 5 \Rightarrow z = 1, 3$
$y = 3 \Rightarrow z = 1$
$y \neq 1$

And we know w is the 1000's digit, and is an odd integer.
And using common sense, we get: $w = 1$. And we have: $w + 9 = y + z$.
So we get: $w = 1 \Rightarrow w + 9 = 10 \Rightarrow y + z = 10$.
And we know that: $y = 9 \Rightarrow z = 1, 3, 5, 7$, and that: $y = 7 \Rightarrow z = 1, 3, 5$.
So we get: $(y, z) = (9, 1)$ or $(7, 3)$.

And thus, the date of birth is Oct 29, 1991 or 1973.

Examples S

0. What is the biggest positive integer?

1. What is the biggest negative integer?

2. What is the smallest negative integer?

3. What is the smallest positive integer?

4. What is the smallest nonzero integer?

5. What is the biggest nonzero integer?

6. What is the biggest 5-digit integer?

7. What is the biggest 5-digit integer where each digit has a different number?

8. What is the smallest 5-digit integer?

9. What is the smallest 5-digit integer where each digit has a different number?

Suggestions or Solutions
To the Problems in the Examples S

0.   What is the biggest positive integer?

A positive integer can be infinitely large. No matter how big positive integer we may come up with, we can still make the larger one. So there is no biggest positive integer.

1.   What is the biggest negative integer?

The smaller the magnitude of a negative number is, the bigger the negative number is. Therefore, the biggest negative integer is -1.

2.   What is the smallest negative integer?

The bigger the magnitude of a negative number is, the smaller the negative number is. A magnitude of an integer can be indefinitely large.
Therefore, there is no smallest negative integer.

3.   What is the smallest positive integer?

The smaller the magnitude of a positive number is, the smaller the positive number is. Therefore, the smallest positive integer is 1.

4.   What is the smallest nonzero integer?

Nonzero integers are negative or positive. Negative integers are smaller than positive integers. We know that there is no smallest negative integer from the solution to the problem 2 above. Therefore, there is no smallest nonzero integer.

5. What is the biggest nonzero integer?

Nonzero integers are negative or positive. Positive integers are bigger than negative integers. We know that there is no biggest positive integer from the solution to the problem 0 above. Therefore, there is no biggest nonzero integer.

6. What is the biggest 5-digit integer?

Suppose that N is a 5-digit integer, and that N = ABCDE where A, B, C, D, and E are the numbers in the digits.
For instance, A is the number in the 10000's digit, and E is the number in the 1's digit.

We know that the integer is positive, and that all numbers need to be the largest in all the digits.
So we can see that A = B = C = D = E = 9 if N is the biggest 5-digit integer.
Therefore, N = 99999.

7. What is the biggest 5-digit integer where each digit has a different number?

Suppose that N is a 5-digit integer, and that N = ABCDE where A, B, C, D, and E are the numbers in the digits.
For instance, A is the number in the 10000's digit, and E is the number in the 1's digit.

We know that

• The integer is positive, and that all numbers need to be the largest in all the digits.

• Each of all the numbers has to be different from all the others.

• The higher digit has to have the larger number than any other number in all the lower digits.

So we can see that A = 9, B = 8, C = 7, D = 6, and E = 5 if N is the biggest 5-digit integer where each digit has a different number. Therefore, N = 98765.

8.   What is the smallest 5-digit integer?

Suppose that N is a 5-digit integer, and that N = ABCDE where A, B, C, D, and E are the numbers in the digits.
For instance, A is the number in the 10000's digit, and E is the number in the 1's digit.

We know that the integer is negative, and that all numbers need to be the largest in magnitude in all the digits.

So we can see that A = B = C = D = E = 9 if N is the biggest 5-digit integer.

Therefore, N = -99999.

What then, is the smallest 5-digit positive integer?

If the N is a 5-digit integer, then A is not zero but the smallest. Then, A = 1.

Also, we can see that all the other 4 digits have to be the smallest, and that they are all 0s.
Therefore, N = 10000

Note that we normally take 0 for a number that has only 1 digit, which is the 1's digit, and in which the number is 0.

9.   What is the smallest 5-digit integer where each digit has a different number?

Suppose that N is a 5-digit integer, and that N = ABCDE where A, B, C, D, and E are the numbers in the digits.

For instance, A is the number in the 10000's digit, and E is the number in the 1's digit.

We know that:

• The integer is negative, and that all numbers need to be the largest in all the digits if the integer is the smallest.

• Each of all the numbers has to be different from all the others.

• The higher digit has to have the larger number than any other number in all the lower digits.

So we can see that A = 9, B = 8, C = 7, D = 6, and E = 5 if N is the smallest 5-digit integer where each digit has a different number.

Therefore, N = -98765.

Examples T

0. What is the smallest integer that is greater than 0?

1. What is the largest integer that is less than 0?

2. What is the largest 6-digit integer?

3. What is the largest 6-digit integer that is less than 0?

4. What is the largest 6-digit integer where 2 digits have 0s?

5. What is the smallest 6-digit integer that is less than 0?

6. What is the smallest 6-digit integer that is greater than 0?

7. What is the smallest 6-digit integer where 2 digits are 0s?

8. What is the largest 6-digit integer where each nonzero digit has a different number and 2 digits have 0s?

9. What is the smallest 6-digit integer where all digits are different?

A. Find the smallest 6-digit integer where 4 digits are different, and 2 digits are 1s.

B. What is the sum of all 8-digit integers?

C. What is the sum of all 8-digit odd integers?

D. What is the sum of all 8-digit even integers?

Suggestions or Solutions
To the Problems in the Examples T

0.   What is the smallest integer that is greater than 0?

It is the smallest positive integer, which is therefore, 1.

1.   What is the largest integer that is less than 0?

It is the largest negative integer, which is thus, -1.

2.   What is the largest 6-digit integer?

Assuming N is the integer we want, we can say that N is positive, and we want every digit in N to be the largest. And we know that the largest is 9. So N is 999999.

3.   What is the largest 6-digit integer that is less than 0?

Assuming N is the integer we want, we can say first, that N is negative. And next, in negative integers, the smaller the digit, the larger the integer. So we want every digit in N to be the smallest. And we know that the smallest is 0. We know however, that the highest digit in N cannot be 0. So N is -100000.

4.   What is the largest 6-digit integer where 2 digits have 0s?

Assuming N is the integer we want, we can say first, that N is positive. And next, in positive integers, the larger the digit, the larger the integer. We know however, two digits in N have to be 0s. So N is 999900.

5.   What is the smallest 6-digit integer that is less than 0?

Assuming N is the integer we want, we can say first, that N is negative.
And next, in negative integers, the larger the digit, the smaller the integer.
So we want every digit in N to be the largest.
And we know that the largest is 9. So N is -999999.

6.   What is the smallest 6-digit integer that is greater than 0?

Assuming N is the integer we want, we can say first, that N is positive.
And next, in positive integers, the smaller the digit, the smaller the integer.
So we want every digit in N to be the smallest. And we know that the smallest is 0.
We know however, that the highest digit in N cannot be 0. So N is 100000.

7.   What is the smallest 6-digit integer where 2 digits are 0s?

Assuming N is the integer we want, we can say first, that N is negative.
And next, in negative integers, the larger the digit, the smaller the integer. We know
however, two digits in N have to be 0s. So N is -999900.

8.   What is the largest 6-digit integer where all nonzero digits are different, and 2 digits are 0s?

Assuming N is the integer we want, we can say first, that N is positive.
And next, in positive integers, the larger the digit, the larger the integer. We know
however, all nonzero digits in N are different, and 2 digits are 0s. So N is 987600.

9. What is the smallest 6-digit integer where all digits are different?

Assuming N is the integer we want, we can say first, that N is negative.
And next, in negative integers, the larger the digit, the smaller the integer. We know
however, all digits in N are different. So N is -987654.

A. Find the smallest 6-digit integer where 4 digits are different, and 2 digits are 1s.

Assuming N is the integer we want, we can say first, that N is negative.
And next, in negative integers, the larger the digit, the smaller the integer. We know however, 4 digits in N are different, two digits in N are 1s. So N is -987611.

B. What is the sum of all 8-digit integers?

The sum is 0.   How come?

Among all the 8-digit integers, the number of ones positive is the same as the number of the ones negative. So the sum is 0.   How come, though?

Assuming P is the sum of all the 8-digit integers positive, we get:
$P = 10000000 + 10000001 + 10000002 + ... + 99999999$.

And assuming N is the sum of all the 8-digit integers negative, we get:
$N = -10000000 - 10000001 - 10000002 - ... - 99999999$
$= -(10000000 + 10000001 + 10000002 + ... + 99999999)$, which is $-P$.

So assuming S is the sum of all the 8-digit integers, we get: $S = P + N = P - P = 0$.

$(10000000 + 10000001 + ... + 99999999) - (10000000 + 10000001 + ... + 99999999)$
$= 10000000 + 10000001 + ... + 99999999 - 10000000 - 10000001 - ... - 99999999$
$= 10000000 - 10000000 + 10000001 - 10000001 + ... + 99999999 - 99999999 = 0$.

C. What is the sum of all 8-digit odd integers?

We know that every 8-digit integer has its negative. For instance, the negative of 23456789 is -23456789. We know that the sum of the two above is 0.
And for any integer odd, we can make a pair of two integers, one is positive, the other is negative, and both have the same absolute value. So the sum of the two in every pair is 0.
And thus, the sum of all the 8-digit odd integers is 0.

D. What is the sum of all 8-digit even integers?

We know that each of all the 8-digit integers has its negative.

For instance, -12345678 is the negative of 12345678, and the negative of -12345678 is 12345678. And in fact, we can make a pair of two numbers for every number, and if one is positive, the other is negative, and both have the same magnitude, that is, the same absolute value. So the sum of the two in each pair is 0.

What then, is the sum of all the 8-digit even integers?

We know we can make a pair of two integers for every integer, and if one is positive, the other is negative, and both have the same magnitude, that is, the same absolute value. So the sum of the two in each pair is 0. That is, the sum of the two in every pair is 0.
So the sum of all the 8-digit even integers is 0, too.

And of course, the sum of all n-digit integers is 0, the sum of all n-digit even integers is 0, and so is the sum of all n-digit odd integers.

Examples U

Factorizing (factoring) an integer, we find first, all its divisors that are prime numbers, called primes, for short, and then, take the product of all the primes. And we call all those primes prime factors. So factorizing an integer, we put the integer in the product of all its prime factors. And normally, if a same prime repeats, we put it in a power, that is, the power of the prime. So for instance, 8 is factorized to 2·2·2, which is therefore, put in 2^3, which is a power of 2, and is called 2 to the third power, or simply, 2 cubed.

0.   Factorize the integers below:

0.   100

1.   2205

2.   126126

3.   42636

4.   525505955

5.   65816751

6.   7807877640

7.   2380749780

8.   980132580713

9.   333028705229

The example below is for those students who took courses on <u>powers and radicals</u>.

1. Put the numbers below in terms of powers of primes:

We have identities (or formulas) on powers, and one of them is as follows:
Assuming n, p, and q are real numbers, we can get: $n^{pq} = (n^p)^q = (n^q)^p$.

For instance, $0.8 = 2^3 10^{-1} = 2^3 (2 \cdot 5)^{-1} = 2^{3-1} 5^{-1} = 2^2 5^{-1}$.

0. 0.032

1. 12.5

2. $\sqrt{22.05}$

3. -1.26126

4. $\sqrt[3]{426.36}$

5. $\sqrt{0.525505955}$

6. $\sqrt[4]{65.816751}$

7. $\sqrt[5]{-98013.2580713}$

2. Put in ascending order, the numbers as follows: 3^{55}, 4^{44}, and 5^{33}.

Suggestions or Solutions
To the Problems in the Examples U

Factorizing (factoring) an integer, we find first, all its divisors that are prime numbers, called primes, for short, and then, take the product of all the primes. And we call all those primes prime factors. So factorizing an integer, we put the integer in the product of all its prime factors. And normally, if a same prime repeats, we put it in a power, that is, the power of the prime. So for instance, 8 is factorized to $2 \cdot 2 \cdot 2$, which is therefore, put in 2^3, which is a power of 2, and is called 2 to the third power, or simply, 2 cubed.

0. $100 = 2 \cdot 50 = 2 \cdot 2 \cdot 25 = 2 \cdot 2 \cdot 5 \cdot 5 = 2^2 \cdot 5^2 = 2^2 5^2$.

1. **2205** $= 5 \cdot 441 = 5 \cdot 3 \cdot 147 = 5 \cdot 3 \cdot 3 \cdot 49 = 3^2 5 \cdot 7^2$

2. **126126** $= 2 \cdot 63063 = 2 \cdot 3 \cdot 21021 = 2 \cdot 3^2 7007 = 2 \cdot 3^2 7 \cdot 1001 = 2 \cdot 3^2 7^2 143 = 2 \cdot 3^2 7^2 11 \cdot 13$

3. **42636** $= 2 \cdot 21318 = 2^2 10659 = 2^2 3 \cdot 3553 = 2^2 3 \cdot 11 \cdot 323 = 2^2 3 \cdot 11 \cdot 17 \cdot 19$

4. **525505955** $= 5 \cdot 105101191 = 5 \cdot 13 \cdot 8084707 = 5 \cdot 13 \cdot 17 \cdot 475571 = 5 \cdot 13 \cdot 17 \cdot 23 \cdot 20677$

$= 5 \cdot 13 \cdot 17 \cdot 23^2 899 = 5 \cdot 13 \cdot 17 \cdot 23^2 29 \cdot 31$.

5. **65816751** $= 3 \cdot 21938917 = 3 \cdot 7 \cdot 3134131 = 3 \cdot 7^2 447733 = 3 \cdot 7^2 11 \cdot 40703$

$= 3 \cdot 7^2 11 \cdot 13 \cdot 3131 = 3 \cdot 7^2 11 \cdot 13 \cdot 31 \cdot 101$.

6. **7807877640** $= 2^2 1951969410 = 2^2 10 \cdot 195196941 = 2^3 5 \cdot 195196941$

$= 2^3 5 \cdot 3 \cdot 65065647 = 2^3 5 \cdot 3^2 21688549 = 2^3 5 \cdot 3^2 17 \cdot 1275797 = 2^3 5 \cdot 3^2 17 \cdot 29 \cdot 43993$

$= 2^3 5 \cdot 3^2 17 \cdot 29^2 1517 = 2^3 5 \cdot 3^2 17 \cdot 29^2 37 \cdot 41$

7. **2380749780** $= 10 \cdot 238074978 = 10 \cdot 2 \cdot 119038489 = 2^2 5 \cdot 3 \cdot 39679163$

$= 2^2 3 \cdot 5 \cdot 19 \cdot 2088377 = 2^2 3 \cdot 5 \cdot 19 \cdot 23 \cdot 90799 = 2^2 3 \cdot 5 \cdot 19 \cdot 23 \cdot 29 \cdot 3131 = 2^2 3 \cdot 5 \cdot 19 \cdot 23 \cdot 29 \cdot 31 \cdot 101$

8. **980132580713** $= 11 \cdot 89102961883 = 11 \cdot 13 \cdot 6854073991 = 11 \cdot 13 \cdot 17 \cdot 403180823$

$= 11 \cdot 13 \cdot 17^2 23716519 = 11 \cdot 13 \cdot 17^2 23 \cdot 1031153 = 11 \cdot 13 \cdot 17^2 23 \cdot 29 \cdot 35557$

$= 11 \cdot 13 \cdot 17^2 23 \cdot 29 \cdot 31 \cdot 1147 = 11 \cdot 13 \cdot 17^2 23 \cdot 29 \cdot 31^2 37$

9. **333028705229** $= 11 \cdot 30275336839 = 11^2 2752303349 = 11^2 17 \cdot 161900197$

$= 11^2 17^2 9523541 = 11^2 17^2 19 \cdot 501239 = 11^2 17^2 19^2 26381 = 11^2 17^2 19^2 23 \cdot 1147$

$= 11^2 17^2 19^2 23 \cdot 31 \cdot 37$

1.   **Put the numbers below in terms of powers of primes:**

Note:

Assuming n, p, and q are real numbers, we can get: $n^{pq} = (n^p)^q = (n^q)^p$.

For instance, $0.8 = 2^3 10^{-1} = 2^3 (2 \cdot 5)^{-1} = 2^{3-1} 5^{-1} = 2^2 5^{-1}$.

0. $0.032 = 32 \cdot 10^{-2} = 4 \cdot 8 \cdot 10^{-2} = 2^2 2^3 10^{-2} = 2^5 10^{-2} = 2^5 (2 \cdot 5)^{-2} = 2^5 2^{-2} 5^{-2} = 2^{5-2} 5^{-2} = 2^3 5^{-2}$

$(= 2^3 / 5^2)$

1.   **12.5**

$125 = 5 \cdot 25 = 5^3$, and thus, $12.5 = 5^3 10^{-1} = 5^3 (2 \cdot 5)^{-1} = 5^3 2^{-1} 5^{-1} = 5^{3-1} 2^{-1} = 5^2 2^{-1} \ (= 5^2 / 2)$

2. $\sqrt{22.05}$

$22.05 = 2205 \cdot 10^{-2}$, and $2205 = 3^2 5 \cdot 7^2$.

So $22.05 = 3^2 5 \cdot 7^2 10^{-2} = 3^2 5 \cdot 7^2 (2 \cdot 5)^{-2} = 3^2 5 \cdot 7^2 2^{-2} 5^{-2} = 2^{-2} 3^2 5^{1-2} 7^2 = 2^{-2} 3^2 5^{-1} 7^2$.

And thus, $\sqrt{22.05} = (2^{-2} 3^2 5^{-1} 7^2)^{1/2} = 2^{-1} 3 \cdot 5^{-1/2} 7 \ (= 3 \cdot 7 / (2 \cdot 5^{1/2}))$

3. -1.26126

$1.26126 = 126126 \cdot 10^{-5}$, and $126126 = 2 \cdot 3^2 7^2 11 \cdot 13$. So we get:

$1.26126 = 2 \cdot 3^2 7^2 11 \cdot 13 \cdot 10^{-5} = 2 \cdot 3^2 7^2 11 \cdot 13 (2 \cdot 5)^{-5} = 2^{1-5} 3^2 5^{-5} 7^2 11 \cdot 13 = 2^{-4} 3^2 5^{-5} 7^2 11 \cdot 13$

$= 3^2 7^2 11 \cdot 13 / (2^4 5^5)$. And thus, $-1.26126 = -3^2 7^2 11 \cdot 13 / (2^4 5^5)$.

4. $\sqrt[3]{426.36}$ To begin with, $426.36 = 42636 \cdot 10^{-2}$, and $42636 = 2^2 3 \cdot 11 \cdot 17 \cdot 19$.

So $426.36 = 2^2 3 \cdot 11 \cdot 17 \cdot 19 \cdot 10^{-2} = 2^2 3 \cdot 11 \cdot 17 \cdot 19 (2 \cdot 5)^{-2} = 2^{2-2} 3 \cdot 5^{-2} 11 \cdot 17 \cdot 19$

$= 2^0 3 \cdot 5^{-2} 11 \cdot 17 \cdot 19 = 1 \cdot 3 \cdot 5^{-2} 11 \cdot 17 \cdot 19 = 3 \cdot 5^{-2} 11 \cdot 17 \cdot 19$.

And thus, $\sqrt[3]{426.36} = (3 \cdot 5^{-2} 11 \cdot 17 \cdot 19)^{1/3} = 3^{1/3} 5^{-2/3} 11^{1/3} 17^{1/3} 19^{1/3} = 3^{1/3} 11^{1/3} 17^{1/3} 19^{1/3} / 5^{2/3}$.

5. $\sqrt{0.525505955}$

$525505955 = 5 \cdot 13 \cdot 17 \cdot 23^2 29 \cdot 31$.

So $0.525505955 = 5 \cdot 13 \cdot 17 \cdot 23^2 29 \cdot 31 \cdot 10^{-1} = 2^{-1} 5^{1-1} 13 \cdot 17 \cdot 23^2 29 \cdot 31 = 2^{-1} \cdot 1 \cdot 13 \cdot 17 \cdot 23^2 29 \cdot 31$

$= 2^{-1} 13 \cdot 17 \cdot 23^2 29 \cdot 31$.

And thus, $\sqrt{0.525505955} = 0.525505955^{1/2} = (2^{-1} 13 \cdot 17 \cdot 23^2 29 \cdot 31)^{1/2}$

$= 2^{-1/2} 13^{1/2} 17^{1/2} 23^{2/2} 29^{1/2} 31^{1/2} = 2^{-1/2} 13^{1/2} 17^{1/2} 23^1 29^{1/2} 31^{1/2} = 13^{1/2} 17^{1/2} 23 \cdot 29^{1/2} 31^{1/2} / 2^{1/2}$.

6. $\sqrt[4]{65.816751}$

$65816751 = 3 \cdot 7^2 11 \cdot 13 \cdot 31 \cdot 101$.

So $65.816751 = 3 \cdot 7^2 11 \cdot 13 \cdot 31 \cdot 101 \cdot 10^{-2} = 2^{-2} 3 \cdot 5^{-2} 7^2 11 \cdot 13 \cdot 31 \cdot 101$.

And thus, $\sqrt[4]{65.816751} = 65.816751^{1/4} = (2^{-2} 3 \cdot 5^{-2} 7^2 11 \cdot 13 \cdot 31 \cdot 101)^{1/4}$

$= 2^{-2/4} 3^{1/4} 5^{-2/4} 7^{2/4} 11^{1/4} 13^{1/4} 31^{1/4} 101^{1/4} = 2^{-1/2} 3^{1/4} 5^{-1/2} 7^{1/2} 11^{1/4} 13^{1/4} 31^{1/4} 101^{1/4}$

$= 3^{1/4} 7^{1/2} 11^{1/4} 13^{1/4} 31^{1/4} 101^{1/4} / (2^{1/2} 5^{1/2})$

7. $\sqrt[5]{-98013.2580713}$

$980132580713 = 11 \cdot 13 \cdot 17^2 23 \cdot 29 \cdot 31^2 37.$

So $98013.2580713 = 11 \cdot 13 \cdot 17^2 23 \cdot 29 \cdot 31^2 37 \cdot 10^{-7} = 2^{-7}5^{-7}11 \cdot 13 \cdot 17^2 23 \cdot 29 \cdot 31^2 37.$

Thus, we get: $98013.2580713^{1/5} = (2^{-7}5^{-7}11 \cdot 13 \cdot 17^2 23 \cdot 29 \cdot 31^2 37)^{1/5}$

$= 2^{-7/5}5^{-7/5}11^{1/5}13^{1/5}17^{2/5}23^{1/5}29^{1/5}31^{2/5}37^{1/5} = 11^{1/5}13^{1/5}17^{2/5}23^{1/5}29^{1/5}31^{2/5}37^{1/5} / (2^{7/5}5^{7/5})$

And thus, $\sqrt[5]{-98013.2580713} = -98013.2580713^{1/5}$

$= -11^{1/5}13^{1/5}17^{2/5}23^{1/5}29^{1/5}31^{2/5}37^{1/5} / 2^{7/5}5^{7/5}.$

And of course, we can put it the way below, too:

$\sqrt[5]{-98013.2580713} = -98013.2580713^{1/5} = -\{11 \cdot 13 \cdot 17^2 23 \cdot 29 \cdot 31^2 37 / (2^7 5^7)\}^{1/5}$ or

$-(11 \cdot 13 \cdot 17^2 23 \cdot 29 \cdot 31^2 37)^{1/5} / (2^7 5^7)^{1/5}.$

2. **Put in ascending order, the numbers as follows: 3^{55}, 4^{44}, and 5^{33}.**

We have: $55 = 11 \cdot 5$, $44 = 11 \cdot 4$, and $33 = 11 \cdot 3$.

So we can get: $3^{55} = (3^5)^{11} = 243^{11}$, $4^{44} = (4^4)^{11} = 256^{11}$, and $5^{33} = (5^3)^{11} = 125^{11}$.

And we know: $125 < 243 < 256$.

Thus, we get: $5^{33} < 3^{55} < 4^{44}$.

Examples V

Note that doing multiplications, we often use a dot · instead of x.

For instance, instead of writing 3 x 5, we just write 3·5. That is, 3 x 5 = 3·5.

And also, doing multiplications using powers as 2^3, we often omit the dot, too, if it is not ambiguous. For instance, we have: 3^2 x 5^2 = $3^2 \cdot 5^2$ =$3^2 5^2$.

Find the GCD (called GCF or GCM, too) and LCM in each of the cases below:

0.   2700 and 2250

1.   25 and 12

2.   7 and 19

3.   8, 9, and 12

4.   24, 45, and 245

5.   60, 75, and 490

6.   4620, 825, and 6930

7.   45045, 225225, and 10395

8.   42042, 315315, 72765, and 231231

9.   58905, 1576575, 20790, 13230, and 114345

Suggestions or Solutions
To the Problems in the Examples V

GCD is the acronym of Greatest Common Divisor, and LCM is the acronym of Least Common Multiple. So the GCD of a set of integers is the largest integer that can divide all the integers in the set, and the LCM is the smallest positive integer that can be divided by all the integers in the set, that is, the smallest multiple common to all the integers.

0.   2700 and 2250

To begin with, factorizing the two integers, we get: $2700 = 2^2 3^3 5^2$, and $2250 = 2 \cdot 3^2 5^3$.

Then, finding the GCD, we take first, the product of all the common factors, which are in this case, common to 2700 and 2250. And all the common factors are 2, 3, and 5. So the product is: $2 \cdot 3 \cdot 5$.

And next, we apply to each common factor the smallest exponent used for the factor.

The smallest exponent applied to the factor 2 is 1, the smallest exponent applied to the factor 3 is 2, and the smallest exponent applied to the factor 5 is 2.

And thus, we get: $GCD = 2 \cdot 3^2 5^2 = 90$.

Next, finding the LCM, we take first, the product of all the prime factors, which are in this case, the prime factors of 2700 and the prime factors of 2250.
And all the prime factors are 2, 3, and 5. So the product is: $2 \cdot 3 \cdot 5$.

And next, we want to apply to each factor the largest exponent used for the factor.

The largest exponent applied to the factor 2 is 2, the largest exponent applied to the factor 3 is 3, and the largest exponent applied to the factor 5 is 3.
And thus, we get: $LCM = 2^2 3^3 5^3 = 13500$.

1. 25 and 12

   GCD = 1, and LCM = $2^2 3 \cdot 5^2 = 300$.

If not sure of how to get it, follow the steps below:

To begin with, factorizing the two integers, we get: $25 = 5^2$, and $12 = 2^2 3$.

Then, finding the GCD, we take first, the product of all the common factors, which are in this case, common to 25 and 12.

In this case though, no factor is common to both numbers.
That is to say that the only divisor common to both numbers is 1.
So the GCD is 1. And in such a case, the two numbers are said to be prime to each other.

Next, finding the LCM, we take first, the product of all the prime factors, which are in this case, the prime factors of 25 and the prime factors of 12.
And all the prime factors are 2, 3, and 5. So the product is: $2 \cdot 3 \cdot 5$.

And next, we apply to each factor the largest exponent used for the factor.

The largest exponent applied to the factor 2 is 2, the largest exponent applied to the factor 3 is 1, and the largest exponent applied to the factor 5 is 2.

And thus, we get: LCM = $2^2 3 \cdot 5^2 = 300$.

2. 7 and 19

   GCD = 1, and LCM = $7 \cdot 19 = 133$.   *If not sure, follow the steps below:*

The two integers are prime integers, and are just called primes, for short.
And no factor is common to primes, which are said to be prime to each other.
That is to say that the only divisor common to both numbers is 1. So the GCD is 1.

And the LCM of primes is just the product of the primes.
So we get: LCM = $7 \cdot 19 = 133$.

3. 8, 9, and 12

To begin with, factorizing the 3 integers, we get: $8 = 2^3$, $9 = 3^2$, and $12 = 2^2 3$.

Then, finding the GCD, we take first, the product of all the common factors, which are in this case, common to 8, 9, and 12. In this case though, no factor is common to all the three numbers, which are thus, prime to each other.
That is to say that the only divisor common to all the three numbers is 1.
So the GCD is 1.

Next, finding the LCM, we take first, the product of all the prime factors, which are in this case, all the prime factors of 8, 9 and 12.
And all the prime factors are 2 and 3. So the product is: $2 \cdot 3$.
And next, we apply to each factor the largest exponent used for the factor.
The largest exponent applied to the factor 2 is 3, and the largest exponent applied to the factor 3 is 2. And thus, we get: LCM $= 2^3 3^2 = 72$.

4. 24, 45, and 245

To begin with, factorizing the 3 integers, we get: $24 = 2^2 3$, $45 = 3^2 5$, and $245 = 5 \cdot 7^2$.
Then, finding the GCD, we take first, the product of all the common factors, which are in this case, common to 24, 45, and 245. In this case though, no factor is common to all the numbers, which are thus, prime to each other.
That is to say that the only divisor common to all the three numbers is 1.
So the GCD is 1.

Next, finding the LCM, we take first, the product of all the prime factors, which are in this case, all the prime factors of 24, 45, and 245.
And all the prime factors are 2, 3, 5, and 7. So the product is: $2 \cdot 3 \cdot 5 \cdot 7$.
And next, we apply to each factor the largest exponent used for the factor.
The largest exponent applied to the factor 2 is 2, the largest exponent applied to the factor 3 is 2, the largest exponent applied to 5 is 1, and the largest exponent applied to 7 is 2. And thus, we get: LCM $= 2^2 3^2 5 \cdot 7^2 = 8820$.

5.   60, 75, and 490

GCD = 5, and LCM = $2^2 3 \cdot 5^2 7^2 = 14700$.

If not sure of how to get it, follow the steps below:

To begin with, factorizing the 3 integers, we get: $60 = 2^2 3 \cdot 5$, $75 = 3 \cdot 5^2$, and $490 = 2 \cdot 5 \cdot 7^2$.

Then, finding the GCD, we take first, the product of all the common factors, which are in this case, common to 60, 75, and 490. And 5 is the only prime factor common.

So the product is: 5.

And next, we apply to each common factor the smallest exponent used for the factor. The smallest exponent applied to the factor 5 is 1.

And thus, we get: GCD = 5.

Next, finding the LCM, we take first, the product of all the prime factors, which are in this case, all the prime factors of 60, 75, and 490. And they are 2, 3, 5, and 7.
So the product is: $2 \cdot 3 \cdot 5 \cdot 7$.

And next, we apply to each factor the largest exponent used for the factor.

The largest exponent applied to the factor 2 is 2, the largest exponent applied to the factor 3 is 1, the largest exponent applied to the factor 5 is 2, and the largest exponent applied to the factor 7 is 2.

And thus, we get: LCM = $2^2 3 \cdot 5^2 7^2 = 14700$.

6.   4620, 825, and 6930

GCD = 3·5·11 = 165, and LCM = $2^2 3^2 5^2 7 \cdot 11$ = 68300.

If not sure of how to get it, follow the steps below:

To begin with, factorizing the 3 integers, we get:
$4620 = 2^2 3 \cdot 5 \cdot 7 \cdot 11$, $825 = 3 \cdot 5^2 11$, and $6390 = 2 \cdot 3^2 5 \cdot 7 \cdot 11$.

Then, finding the GCD, we take first, the product of all the common factors, which are in this case, common to 4620, 825, and 6930.

And all the common factors are 3, 5, and 11. So the product is: 3·5·11.

And next, we apply to each common factor the smallest exponent used for the factor.

The smallest exponent applied to the factor 3 is 1, the smallest exponent applied to the factor 5 is 1, and the smallest exponent applied to the factor 11 is 1.

And thus, we get: GCD = 3·5·11 = 165.

Next, finding the LCM, we take first, the product of all the prime factors, which are in this case, all the prime factors of 4620, 825, and 6930.

And all the prime factors are 2, 3, 5, 7, and 11. So the product is: 2·3·5·7·11.

And next, we want to apply to each factor the largest exponent used for the factor.

The largest exponent applied to the factor 2 is 2, the largest exponent applied to the factor 3 is 2, the largest exponent applied to the factor 5 is 2, the largest exponent applied to the factor 7 is 1, and the largest exponent applied to the factor 11 is 1.

And thus, we get: LCM = $2^2 3^2 5^2 7 \cdot 11$ = 68300.

7.   45045, 225225, and 10395

GCD = $3^2 \cdot 5 \cdot 7 \cdot 11 = 3465$, and LCM = $3^3 5^2 7 \cdot 11 \cdot 13 = 675675$.

If not sure of how to get it, follow the steps below:

To begin with, factorizing the 3 integers, we get:

$44045 = 3^2 5 \cdot 7 \cdot 11 \cdot 13$, $225225 = 3^2 5^2 7 \cdot 11 \cdot 13$, and $10395 = 3^3 5 \cdot 7 \cdot 11$.

Then, finding the GCD, we take first, the product of all the common factors, which are in this case, common to 45045, 225225, and 10395.

And all the common factors are 3, 5, 7, and 11. So the product is: $3 \cdot 5 \cdot 7 \cdot 11$.

And next, we apply to each common factor the smallest exponent used for the factor.

The smallest exponent applied to the factor 3 is 2, the smallest exponent applied to the factor 5 is 1, the smallest exponent applied to the factor 7 is 1, and the smallest exponent applied to the factor 11 is 1.

And thus, we get: GCD = $3^2 \cdot 5 \cdot 7 \cdot 11 = 3465$.

Next, finding the LCM, we take first, the product of all the prime factors, which are in this case, all the prime factors of 45045, 225225, and 10395.

And all the prime factors are 3, 5, 7, 11, and 13. So the product is: $3 \cdot 5 \cdot 7 \cdot 11 \cdot 13$.

And next, we want to apply to each factor the largest exponent used for the factor.

The largest exponent applied to the factor 3 is 3, the largest exponent applied to the factor 5 is 2, the largest exponent applied to the factor 7 is 1, the largest exponent applied to the factor 11 is 1, and the largest exponent applied to the factor 13 is 1.

And thus, we get: LCM = $3^3 5^2 7 \cdot 11 \cdot 13 = 675675$.

8.   42042, 315315, 72765, and 231231

GCD = $3 \cdot 7^2 11 = 1617$, and LCM = $2 \cdot 3^3 5 \cdot 7^2 11^2 13 = 1891890$.

If not sure of how to get it, follow the steps below:

To begin with, factorizing the 3 integers, we get:

$42042 = 2 \cdot 3 \cdot 7^2 11 \cdot 13$, $315315 = 3^2 5 \cdot 7^2 11 \cdot 13$, $72765 = 3^3 5 \cdot 7^2 11$, and $231231 = 3 \cdot 7^2 11^2 13$.

Then, finding the GCD, we take first, the product of all the common factors, which are in this case, common to 42042, 315315, 72765, and 231231.

And all the common factors are 3, 7, and 11. So the product is: $3 \cdot 7 \cdot 11$.

And next, we apply to each common factor the smallest exponent used for the factor.

The smallest exponent applied to the factor 3 is 1, the smallest exponent applied to the factor 7 is 2, and the smallest exponent applied to the factor 11 is 1.

And thus, we get: GCD = $3 \cdot 7^2 11 = 1617$.

Next, finding the LCM, we take first, the product of all the prime factors, which are in this case, all the prime factors of 42042, 315315, 72765, and 231231.

And all the prime factors are 2, 3, 5, 7, 11, and 13. So the product is: $2 \cdot 3 \cdot 5 \cdot 7 \cdot 11 \cdot 13$.

And next, we want to apply to each factor the largest exponent used for the factor.

The largest exponent applied to the factor 2 is 1, the largest exponent applied to the factor 3 is 3, the largest exponent applied to the factor 5 is 1, the largest exponent applied to the factor 7 is 2, the largest exponent applied to the factor 11 is 2, and the largest exponent applied to the factor 13 is 1.

And thus, we get: LCM = $2 \cdot 3^3 5 \cdot 7^2 11^2 13 = 1891890$.

9. **58905, 1576575, 20790, 13230, and 114345**

GCD $= 3^2 5 \cdot 7 = 315$, and LCM $= 2 \cdot 3^3 5^2 7^2 11^2 13 \cdot 17 = 1768917150$.

If not sure of how to get it, follow the steps below:

To begin with, factorizing the 3 integers, we get:

$58905 = 3^2 5 \cdot 7 \cdot 11 \cdot 17$, $1576575 = 3^2 5^2 7^2 11 \cdot 13$, $20790 = 2 \cdot 3^3 5 \cdot 7 \cdot 11$, $13230 = 2 \cdot 3^3 5 \cdot 7^2$, and $114345 = 3^3 5 \cdot 7 \cdot 11^2$.

Then, finding the GCD, we take first, the product of all the common factors, which are in this case, common to 58905, 1576575, 20790, 13230, and 114345.

And all the common factors are 3, 5, and 7. So the product is: $3 \cdot 5 \cdot 7$.

And next, we apply to each common factor the smallest exponent used for the factor.

The smallest exponent applied to the factor 3 is 2, the smallest exponent applied to the factor 5 is 1, and the smallest exponent applied to the factor 7 is 1.

And thus, we get: GCD $= 3^2 5 \cdot 7 = 315$.

Next, finding the LCM, we take first, the product of all the prime factors, which are in this case, all the prime factors of 58905, 1576575, 20790, 13230, and 114345.

And all the prime factors are 2, 3, 5, 7, 11, 13, and 17.
So the product is: $2 \cdot 3 \cdot 5 \cdot 7 \cdot 11 \cdot 13 \cdot 17$.

And next, we want to apply to each factor the largest exponent used for the factor.

The largest exponent applied to the factor 2 is 1, the largest exponent applied to the factor 3 is 3, the largest exponent applied to the factor 5 is 2, the largest exponent applied to the factor 7 is 2, the largest exponent applied to the factor 11 is 2, the largest exponent applied to the factor 13 is 1, and the largest exponent applied to the factor 17 is 1.

And thus, we get: LCM $= 2 \cdot 3^3 5^2 7^2 11^2 13 \cdot 17 = 1768917150$.

Examples W

0.   Assuming that a is an integer positive, and that if we divide it by 7, the quotient is 5, and the remainder is 6, find a.

1.   Assuming a is a positive integer, and if we divide it by 7, the remainder is 6, find a.

2.   How many two-digit positive integers can we divide by 3 and 5?

3.   Find the smallest integer a satisfying $b = \frac{4}{5}a$ where b is a positive integer.

4.   Find the largest integer a satisfying $b = \frac{2}{3}a$ where b is a negative integer.

5.   Assuming the GCD (GCF) of five integers is 24, find all the common divisors of the integers.

6.   Assuming m is between 0 and 1000, and is a common multiple of 4, 12, and 15, find the number of all the numbers that can be m.

7.   Assuming $a = \frac{8}{5} \cdot \frac{x}{y}$, and $b = \frac{70}{3} \cdot \frac{x}{y}$ where a and b are positive integers, find x and y that are positive integers and prime to each other.

Suggestions or Solutions
To the Examples W

0. Assuming that *a* is an integer positive, and that if we divide it by 7, the quotient is 5, and the remainder is 6, find *a*.

$a = 41.$ *If not sure of how to get it, follow the steps below:*

First, dividing *a* by 7, we get: $\dfrac{a}{7}$.

Next, if the quotient is 5 and the remainder is 6, which of the two below do we get?

$\dfrac{a}{7} = 5 + 6$ and $\dfrac{a}{7} = 5 + \dfrac{6}{7}$

We get: $\dfrac{a}{7} = 5 + \dfrac{6}{7}$. So we get: $a = 5\cdot 7 + 6 = 41$.

1. Assuming *a* is a positive integer, and if we divide it by 55, the remainder is 7, find *a*.

$a = 55q + 7$, where *q* is a nonnegative integer.

If not sure of how to get it, follow the steps below:

First, dividing *a* by 55, we get: $\dfrac{a}{55}$.

Next, if the quotient is *q* and the remainder is 7, which of the two below do we get?

$\dfrac{a}{55} = q + 7$ and $\dfrac{a}{55} = q + \dfrac{7}{55}$.

We get: $\dfrac{a}{55} = q + \dfrac{7}{55}$.   What then, is q? Which one of the three below?

A. q is an integer.   B. q is a positive integer.   C. q is a nonnegative integer.

q is one of 0, 1, 2, 3, …, and thus, is one of nonnegative integers.

So we get: $a = 55q + 7$, where q is a nonnegative integer.

That is, a can be any of $55 \cdot 0 + 7 = 7$, $55 + 7 = 62$, $55 \cdot 2 + 7 = 117$, and so forth.

2.   How many 3-digit positive integers can we divide by 4 and 6?

75.   *If not sure of how to get it, follow the steps below:*

First, what integers can 4 and 6 divide?

They are common multiples of 4 and 6. So 4 and 6 both can divide a common multiple of 4 and 6. How large then, can it be?

It can be infinitely large. How small then, can it be?

It can be as small as not 24 but 12, which is therefore, the smallest common multiple. And we often call it the least common multiple, and usually call it the LCM, which is the acronym of the **l**east **c**ommon **m**ultiple.

What then, are numbers that 4 and 6 can divide?

They are multiples of the LCM of the two integers 4 and 6. And the LCM is 12.

So they are multiples of 12. And assuming m is such a multiple, we can put it this way: $m = 12n$ where n is an integer.

Now, we know that the numbers we want are 3-digit integers positive.

And of all the 3-digit positive integers, the smallest is 100, and the largest is 999.

How then, can we get the number of 3-digit positive integers that 12 can divide?

We can get it finding n that satisfies the relation as follows: $\mathbf{100 \leq 12n \leq 999}$.

And it is the value of n, of course. So finding n, we get:

$$100 \leq 12n \leq 999 \Rightarrow \frac{100}{12} \leq n \leq \frac{999}{12} \Rightarrow 8 + \frac{4}{12} \leq n \leq \frac{333}{4}$$

$$\Rightarrow 8 + \frac{1}{3} \leq n \leq 83 + \frac{1}{4} \Rightarrow 9 \leq n \leq 83.$$

That is to say that n can be any integer between 8 and 84.

In other words, n can be any of all the integers from 9 to 83.   How com though?

We know $\mathbf{12n}$ has to be a 3-digit positive integer. And we have: $12 \cdot 8 = 96$, which is just a 2-digit integer, $12 \cdot 9 = 108$, which is a 3-digit integer, $12 \cdot 83 = 996$, which is a 3-digit integer, and $12 \cdot 84 = 1008$, which is a 4-digit integer.

And thus, the number of all 3-digit positive integers 3 and 6 divide is: $83 - 9 + 1 = 75$.

3.   **Find the smallest integer a satisfying $b = \frac{4}{5}a$ where b is a positive integer.**

We have: $b = \frac{4}{5}a$. And we can put it this way, too: $\dfrac{b}{a} = \dfrac{4}{5}$.

So the ratio of b to a is $\frac{4}{5}$. In other words, $b : a = 4 : 5$.

So the smallest integer a is 5, since b is a positive integer.

4. **Find the largest integer a satisfying $b = \frac{2}{3}a$ where b is a negative integer.**

We have: $b = \frac{2}{3}a$. And we can put it this way, too: $\dfrac{b}{a} = \dfrac{2}{3}$.

So the ratio of b to a is $\frac{2}{3}$. In other words, $b : a = 2 : 3$. So the largest integer a is -3, since b is a negative integer. And in this case, $b = \textbf{-2}$, of course.

5. **Assuming the GCD (GCF) of five integers is 24, find all the common divisors of the integers.**

> 1, 2, 8, 3, 6, 12, and 24.   *If not sure of how to get it, follow the steps below:*

We know that the GCD of a set of integers is the greatest common divisor of all the integers in the set. And we know that a factor is a divisor. So GCD is called GCF, too, which is the acronym of Greatest Common Factor.

And thus, the GCD can divide all the integers in the set.
So if an integer can divide the GCD, the integer can divide all the integers in the set.
That is to say that a divisor of the GCD can divide all the integers in the set.
In other words, a divisor of the GCD is a common divisor of all the integers in the set.
And thus, all the divisors of the GCD can divide all the integers in the set.
That is to say that all the divisors of the GCD are all the common divisors of all the integers in the set.

Now, the GCD of the five integers is the greatest common divisor of all the five integers. And thus, the GCD can divide all the integers in the set. We know that the GCD is 24.
So if an integer can divide the GCD 24, the integer can divide all the five integers.
That is to say that a divisor of 24 can divide all the five integers.
And thus, all the divisors of 24 can divide all the five integers.

So first, factorizing the GCD, we get: $24 = 8 \cdot 3 = 2^3 \cdot 3$.
Thus next, finding all the divisors of 24, we get:
$1, 2, 2^2 = 4, 2^3 = 8, 3, 2 \cdot 3 = 6, 2^2 \cdot 3 = 12,$ and $2^3 \cdot 3 = 24$.

6. Assuming *m* is between 0 and 1000, and is a common multiple of 4, 12, and 15, find the number of all the numbers that can be *m*.

16. *If not sure of how to get it, follow the steps below:*

We know that the LCM of a set of integers is the least common multiple of all the integers in the set.

So all the multiples of the LCM are common multiples of all the integers in the set.

Now, in this case, all the integers in the set are 4, 12, and 15.

And finding the LCM, we factorize the three integers first.

So factorizing them, we get: $4 = 2^2$, $12 = 4 \cdot 3 = 2^2 3$, and $15 = 3 \cdot 5$.

And next, we put all the factors in a product form.
So getting the product, we get: $2 \cdot 3 \cdot 5$.

And next, finding the LCM, we apply to each factor the largest exponent used for the factor. Thus, we get: $LCM = 2^2 3 \cdot 5 = 60$.

Now, we know *m* is a multiple of the LCM.

So we can set: *m* = **60k** where *k* is an integer.

And we want *m* to be between 0 and 1000.

So we get: $0 < 60k < 1000 \Rightarrow 0 < k < \frac{1000}{60} = \frac{100}{6} = \frac{50}{3} = 16 + \frac{2}{3} \Rightarrow 1 \le k \le 16$.

And thus, the number of all the numbers that can be m is: $16 - 1 + 1 = 16$.

7. Assuming $a = \dfrac{8}{5} \cdot \dfrac{x}{y}$, and $b = \dfrac{70}{3} \cdot \dfrac{x}{y}$ where a and b are positive integers, find x and y that are positive integers and prime to each other.

$(x, y) = (15k, 1)$ where k is a positive integer.

$(x, y) = (30m + 15, 2)$ where m is a nonnegative integer, that is, $m = 0$, 1, 2, etc.

If not sure of how to get it, follow the steps below:

First, we know a is a positive integer.
So we can say that x is a multiple of 5, and that y is a divisor of 8.

Next, b is a positive integer, too. So we can say that x is a multiple of 3, too, and that y is a divisor of 70, also. What then, can we say about x and y?

We can say: x is a common multiple of 5 and 3, and y is a common divisor of 8 and 70. What then, can we say about x and y?

We know that the smallest common multiple is the LCM of 5 and 3, and that greatest divisor is the GCD of 8 and 70.

So we can say that x is a multiple of the LCM of 5 and 3, and that y is a divisor of the GCD of 8 and 70. And thus, we want to get the LCM and the GCD.

So first, the LCM of 5 and 3 is 15.

And next, finding the GCD, we factorize 8 and 70 first.
So factorizing them, we get: $8 = 2^3$, and $70 = 2 \cdot 5 \cdot 7$.

Next, we put all common factors in a product form.
In this case, 2 is the only common factor.

Next, we apply to each factor the smallest exponent used for the factor. Thus, we get: the GCD = 2.

Now, we know x is a multiple of the LCM, which is 15.

So we can set: $x = 15k$ where k is a positive integer.

Next, we know y is a divisor of the GCD, which is 2. So we get: $y = 2$ or 1.

That's not it though.   What then, is the next?

We know x and y are prime to each other. So we can put x and y in the two cases below:

- $(x, y) = (15k, 1)$ where k is a positive integer.
So for instance, $x = 15$ and $y = 1$, $x = 30$ and $y = 1$, $x = 45$ and $y = 1$, and so forth.

- $(x, y) = (15n, 2)$ where n is an odd positive integer.
So for instance, $x = 15$ and $y = 2$, $x = 45$ and $y = 2$, $x = 75$ and $y = 2$, and so forth.

And since n is an odd positive integer, we can put $15n$ the way below, too:

$15n = 15(2m + 1) = 30m + 15$ where m is a nonnegative integer

So we can put the two above the way below, too:

- $(x, y) = (15k, 1)$ where k is a positive integer.
- $(x, y) = (30m + 15, 2)$ where m is a nonnegative integer, that is, $m = 0, 1, 2$, etc.

Examples X

0.   Assuming that a is a positive integer, and that dividing 40 by a, we get 8 as a remainder, and dividing 50 by a, we get 2 as a remainder, find a.

1.   Assuming a and b are integers positive, $\dfrac{a}{b} = \dfrac{12}{5}$, and the sum of their GCD (GCF) and LCM is 305, find a and b.

2.   Assuming a is a positive integer, GCD of a and 32 is 8, and LCM is 160, find a.

3.   Assuming we have 60 apples and 90 bananas, find the maximum number of people to whom we can give the same number of apples and the same number of bananas.

4.   Assuming we have 60 apples, 90 bananas, 120 pears, and 150 oranges, find the maximum number of people to whom we can give the same number of each fruit evenly. Assuming for instance, we have 3 apples and 6 bananas, we can give 1 apple and 2 bananas to 3 people evenly, and thus, the maximum number of people is 3.

5.   Suppose we have 27,225 apples, 370,260 pears, and 1,262,250 bananas. Suppose also, we want to deliver the fruits to a number of stores, but each store needs to get as many of each fruit as possible. And each of the stores needs to get the same amount for each fruit. For instance, if one store gets 5 apples, each of all the others has to get 5 apples each, too.

How many of each fruit then, do we need to deliver to each store maximizing the number of stores?

And how many of each fruit, do we need to deliver to each store minimizing the number of stores? The minimum is greater than 1, of course.

Suggestions or Solutions
To the Examples X

0.   Assuming that a is a positive integer, and that dividing 40 by a, we get 8 as a remainder, and dividing 50 by a, we get 2 as a remainder, find a.

$a = 16$.   *If not sure of how to get it, follow the steps below:*

To begin with, dividing 40 by a, we get 8 as a remainder.

So we get: $40 = ka + 8$ where k is an integer, and $a > 8$ since the remainder is less than a.

Thus, we get: $ka = 40 - 8 = 32 \Rightarrow ka = 32$.

So we can say that a divides 32. That is to say that a is a divisor of 32.

Next, dividing 50 by a, we get 2 as a remainder.

So we get: $50 = qa + 2$ where q is an integer, and $a > 2$ since the remainder is less than a.

Thus, we get: $qa = 50 - 2 = 48 \Rightarrow qa = 48$.

So we can say that a divides 48, too. That is, a is a divisor of 48.

What then, can we say about a?

We know a divides 32 as well as 48.

So we can say that a is a common divisor of 32 and 48.

And the GCD of 32 and 48 is the greatest common divisor of 32 and 48.

That is to say that of all divisors common to 32 and 48, the GCD is the largest.

What then, a can be?

We can say that a is one of divisors of the GCD.

That's because a divisor of the GCD is a common divisor of 32 and 48, too.

In other words, a factor of the GCD is a factor common to 32 and 48, too.

So let's now find the GCD.

$32 = 2^5$, and $48 = 2 \cdot 24 = 6 \cdot 8 = 3 \cdot 16 = 2^4 3$. So the GCD is 2^4, which is 16.

And thus, a can be one of 1, 2, 4, 8, and 16, which are divisors of 16.
And we know: $a > 8$. So we get: $a = 16$.

9. **Assuming a and b are integers positive, $\dfrac{a}{b} = \dfrac{12}{5}$, and the sum of their GCD (GCF) and LCM is 305, find a and b.**

$a = 60$, and $b = 25$.   *If not sure of how to get it, follow the steps below:*

Assuming first, G is the GCD of a and b, we can set: $a = 12G$, and $b = 5G$.

It's because 12 and 5 are prime to each other, and if they are not, G is not the GCD of a and b, either.

And since G is the largest divisor common to a and b, we get: $\frac{a}{b} = \frac{12G}{5G} = \frac{12}{5}$.

And assuming next, L is the LCM of a and b, we can set: $L = 12 \cdot 5G = 60G$.

That's because 12 and 5 are prime to each other, $a = 12G$, $b = 5G$, and L is the smallest multiple common to a and b, and thus, L is the smallest that a and b both can divide.

And next, we have: $G + L = 305$.   What then, do we get?

We have: $L = 60G$, too.

So we get: $G + L = G + 60G = 61G = 305 \Rightarrow G = 5$.

Thus, we get: $a = 12G = 60$, and $b = 5G = 25$.

94

A.   Assuming *a* is a positive integer, GCD of *a* and 32 is 8, and LCM is 160, find *a*.

a = 40.   *If not sure of how to get it, follow the steps below:*

Assuming first, **G** is the GCD of *a* and 32, we can set: **a = kG**, where **k** is an integer.

Next, we have: **32 = 4 x 8 = 4G**, since **G = 8**.

How then, can we put the LCM of *a* and 32 in terms of **G**?

We have: **a = kG**, and **32 = 4G**, where **G** is the GCD of *a* and 32.

So we can say that **k** and 4 are prime to each other.
And thus, assuming **L** is the LCM of *a* and 32, we can set: **L = 4kG**.

And we know: **L = 160**, and **G = 8**.

So we get: $L = 4kG \Rightarrow 160 = 4{\cdot}k{\cdot}8 = 32k \Rightarrow k = 160/32 = 5$.

And we know: **a = kG**, where **k = 5**, and **G = 8**. So we get: **a = 40**.

B.   Assuming we have 60 apples and 90 bananas, find the maximum number of people to whom we can give the same number of apples and the same number of bananas.

30.   *If not sure of how to get it, follow the steps below:*

Suppose in a division, the divisor is a positive integer.
Then, the larger the divisor, the larger the number of equal pieces.
Dividing for instance, 1 by 3, we divide 1 into three equal pieces. And each piece is: $\frac{1}{3}$.

And dividing 10 by 5, we divide 10 into 5 equal parts. And each part is 2.

So for instance, giving apples to people, and dividing the number of apples by the number of people, we get the number of apples each person gets. And the larger the number of people, the larger the divisor, because the divisor is the number of people.

And the same is true for bananas, too.

So the maximum number of people in this problem is the greatest common divisor of the number of apples and the number of bananas.
In short, the maximum number is the GCD of 60 and 90.

So first, factorizing 60 and 90, we get:
$60 = 2 \cdot 30 = 2 \cdot 2 \cdot 15 = 2^2 3 \cdot 5$, and $90 = 2 \cdot 45 = 2 \cdot 5 \cdot 9 = 2 \cdot 3^2 5$.

Thus next, the GCD $= 2 \cdot 3 \cdot 5 = 30$, which is the maximum number of people.

And we have: $60 = 2 \cdot 30$, and $90 = 3 \cdot 30$.
So 30 people can get 2 apples each and 3 bananas each.

This is one of the uses of GCD (the greatest common divisor).

The GCD of a set of amounts can indicate the largest number of equal parts each of all the amounts can be divided into.

So in this case, we can divide 60 apples into up to 30 equal parts, each part has 2 apples, and we can divide 90 bananas into 30 equal parts, and each part has 3 bananas.

C.   Assuming we have 60 apples, 90 bananas, 120 pears, and 150 oranges, find the maximum number of people to whom we can give the same number of each fruit evenly. Assuming for instance, we have 3 apples and 6 bananas, we can give 1 apple and 2 bananas to 3 people evenly, and thus, the maximum number of people is 3.

30.   *If not sure of how to get it, follow the steps below:*

This is no other than the problem **B** above.

So the maximum number of people is the greatest common divisor of the four numbers of the fruits.

In short, the maximum number is the GCD of 60, 90, 120, and 150.

So first, factorizing 60, 90, 120, and 150, we get:
$60 = 2^2 3 \cdot 5$, $90 = 2 \cdot 3^2 5$, $120 = 2^3 3 \cdot 5$, and $150 = 2 \cdot 3 \cdot 5^2$.

Thus next, the GCD $= 2 \cdot 3 \cdot 5 = 30$, which is the maximum number of people.
And we have: $60 = 2 \cdot 30$, and $90 = 3 \cdot 30$.

So 30 people can get 2 apples each and 3 bananas each.
The maximum number of people is the greatest common divisor of the number of apples and the number of bananas.

In short, the maximum number is the GCD of 60 and 90.

So first, factorizing 60 and 90, we get:

$60 = 2 \cdot 30 = 2 \cdot 2 \cdot 15 = 2^2 3 \cdot 5$, and $90 = 2 \cdot 45 = 2 \cdot 5 \cdot 9 = 2 \cdot 3^2 5$.

Thus next, the GCD $= 2 \cdot 3 \cdot 5 = 30$, which is the maximum number of people.

And we have: $60 = 2 \cdot 30$, $90 = 3 \cdot 30$, $120 = 4 \cdot 30$, and $150 = 5 \cdot 30$.

So 30 people can get 2 apples each, 3 bananas each, 4 pears each, and 5 oranges each.

And notice that 2, 3, 4, and 5 are prime to each other.
That is, other than 1, the four integers have no common divisor.
And saying no common divisor, we mean the GCD is 1.

D.   Suppose we have 27,225 apples, 370,260 pears, and 1,262,250 bananas. Suppose also, we want to deliver the fruits to a number of stores, but each store needs to get as many of each fruit as possible. And each of the stores needs to get the same amount for each fruit. For instance, if one store gets 5 apples, each of all the others has to get 5 apples each, too.

How many of each fruit then, do we need to deliver to each store maximizing the number of stores?

And how many of each fruit, do we need to deliver to each store minimizing the number of stores? The minimum is greater than 1, of course.

The first question is no other than the problem **C** above.
So the maximum number of stores is the greatest common divisor of the 3 numbers of the fruits.
In short, the maximum number is the GCD of 27225, 370260, and 1262250.

So first, factorizing the 3 numbers, we get:

$27225 = 3^2 5^2 11^2$, $370260 = 2^2 3^2 5 \cdot 11^2 17$, and $1262250 = 2 \cdot 3^3 \cdot 5^3 11 \cdot 17$.

Thus next, the GCD $= 3^2 \cdot 5 \cdot 11 = 495$, which is the maximum number of stores.
And assuming $G = $ **GCD**, we have:

$27225 = 11G$, $370260 = 2^2 \cdot 11 \cdot 17G$, and $1262250 = 2 \cdot 3 \cdot 5^2 \cdot 17G$.

So 495 stores can get 11 apples each, $2^2 \cdot 11 \cdot 17$ pears each, and $2 \cdot 3 \cdot 5^2 \cdot 17$ bananas each.

And notice that 11, $2^2 \cdot 11 \cdot 17$, and $2 \cdot 3 \cdot 5^2 \cdot 17$ are prime to each other.
That is, no factor is common to all the three numbers.

And next, we want to find the number of each fruit we need to deliver to each store minimizing the number of stores.

What then, do we need to find for the three numbers 27225, 370260, and 1262250?

We want to find not the greatest but the smallest common divisor.
What then, is the smallest common divisor?

We have: $27225 = 3^2 5^2 11^2$, $370260 = 2^2 3^2 5 \cdot 11^2 17$, and $1262250 = 2 \cdot 3^3 \cdot 5^3 11 \cdot 17$.

So we can see that 3 is the divisor the smallest common to all the three numbers. And we can put the three numbers the way below:

$27225 = 3 \cdot 3 \cdot 5^2 11^2$, $370260 = 3 \cdot 2^2 3 \cdot 5 \cdot 11^2 17$, and $1262250 = 3 \cdot 2 \cdot 3^2 \cdot 5^3 11 \cdot 17$.

So 3 stores can get $3 \cdot 5^2 11^2$ apples each, $2^2 3 \cdot 5 \cdot 11^2 17$ pears each, and $2 \cdot 3^2 \cdot 5^3 11 \cdot 17$ bananas each.

Examples Y

0. Show that $n^5 - n$ is a multiple of 30 if $|n| > 1$ is an integer.

1. Suppose a, b, and c are positive integers, and $p = a^2 + b^2 + c^2$. Then:

A. Show that if a is not a multiple of 3, the remainder in the division of a^2 by 3 is 1.
B. Show that if none of a, b, and c is a multiple of 3, p is a multiple of 3.
C. Show that if a, b, c, and p are primes, at least one of a, b, and c is 3.

2. Assuming a is an integer positive, and $n(a)$ is the number of all positive divisors of a, find the minimum of x for which $n(108)n(32)n(x) = 5400$.

3. Assuming a and b are positive integers, find the smallest a and b for which $a^2 = 2b^2$.

4. Assuming x, y, and z are integers, find the smallest y for which we get: $x^2 + z^2 = y^3$ if $x > y > z > 0$.

5. Put 2003 in the sum of powers of 2. Putting for instance, 83 in the sum of powers of 2, we get: $83 = 2^6 + 2^4 + 2^1 + 2^0$.

6. Put each of 135 and 598 in terms of the powers of digits in the number itself. For instance, $153 = 1 + 5^3 + 3^3$, $24 = 2^3 + 4^2$, and $2427 = 2 + 4^2 + 2^3 + 7^4$.

7. Show that if x is a positive integer and has an odd number of divisors, there is an integer n for which we get: $x = n^2$.

Suggestions or Solutions
To the Examples Y

0.   Show that $n^5 - n$ is a multiple of 30 if $|n| > 1$ is an integer.

To begin with, 5 divides 30, and so does 6.

So 30 is a common multiple of 5 and 6. What then, about a multiple of 30?

It is a common multiple of 5 and 6.

And thus, if $(n^5 - n)$ is a common multiple of 5 and 6, it is a multiple of 30.

How do we know then, if $(n^5 - n)$ is a common multiple of 5 and 6?

We can have: $n^5 - n = n(n^4 - 1) = n(n^2 - 1)(n^2 + 1) = (n - 1)n(n + 1)(n^2 + 1)$. So what?

We know $(n - 1)n(n + 1)$ is a product of three contiguous integers, since n is an integer.

And a multiple of three contiguous integers is a multiple of 6.

That's because one the three integers is even, and one of the others is a multiple of 3.

For instance, $12 \cdot 13 \cdot 14 = 3 \cdot 4 \cdot 13 \cdot 2 \cdot 7 = 6 \cdot 4 \cdot 13 \cdot 7$. So $(n^5 - n)$ is a multiple of 6.

And next, we can put all integers the way below: $5k$, $5k \pm 1$, and $5k \pm 2$, where k is an integer. For instance, if $k = -1$, we get: -7, -6, -5, -4, -3, if $k = 0$, we get: -2, -1, 0, 1, 2, and if $k = 1$, we get: 3, 4, 5, 6, 7.

Then, first, if $n = 5k$, we can say, of course, n is a multiple of 5.

Next, if $n = 5k \pm 1$, we get: $n \pm 1 = 5k$. That is, $n + 1 = 5k$ or $n - 1 = 5k$.

So $(n - 1)$ or $(n + 1)$ is a multiple of 5.

And next, if $n = 5k \pm 2$, we get: $n^2 + 1 = (5k \pm 2)^2 + 1 = 25k^2 \pm 20k + 5$.
So we can say that $(n^2 + 1)$ is a multiple of 5.

And thus, $(n-1)n(n+1)(n^2+1)$, that is, $(n^5 - n)$ is a multiple of 5, too.
So $(n^5 - n)$ is a common multiple of 5 and 6, and thus, is a multiple of 30.

1. **Suppose a, b, and c are positive integers, and $p = a^2 + b^2 + c^2$. Then:**

A. **Show that if a is not a multiple of 3, the remainder in the division of a^2 by 3 is 1.**
B. **Show that if none of a, b, and c is a multiple of 3, p is a multiple of 3.**
C. **Show that if a, b, c, and p are primes, at least one of a, b, and c is 3.**

A.

To begin with, if a is not a multiple of 3, we can put it this way: $a = 3k + 1$ or $3k + 2$, where k is an integer ≥ 0. And more simply, we can put it this way, too: $a = 3k \pm 1$.

Then, we get: $a^2 = (3k \pm 1)^2 = 9k^2 \pm 6k + 1 = 3(3k^2 \pm 2k) + 1$.
So the remainder in the division of a^2 by 3 is 1.

B.

To begin with, we know that if a is not a multiple of 3, the remainder in the division of a^2 by 3 is 1.
And the same is true, too, for b^2, and c^2, because neither of b and c is a multiple of 3.
So the remainder in the division of b^2 by 3 is 1.
And the remainder in the division of c^2 by 3 is 1, too.

Thus next, setting: $a^2 = 3A + 1$, $b^2 = 3B + 1$, and $c^2 = 3C + 1$ where A, B, and C are positive integers, we get: $p = a^2 + b^2 + c^2 = 3(A + B + C) + 3$.
So we can say that p is a multiple of 3.

C.
Suppose that none of a, b, and c is 3.

Then, none of them is a multiple of 3, since they are primes. We know from the part B, p is a multiple of 3. So we get: $p = 3$, because p is a prime.
However, we get: $p = a^2 + b^2 + c^2 \geq 12$, because $a \geq 2$, $b \geq 2$, and $c \geq 2$, since a, b, and c are primes.

That is to say that it can be the case where $p = 12$, which contradicts however, the fact that p is a multiple of 3. So at least one of a, b and c is 3.

2.   Assuming a is an integer positive, and $n(a)$ is the number of all positive divisors of a, find the minimum of x for which $n(108)n(32)n(x) = 5400$.

To begin with, $n(a)$ is the number of all positive divisors of a. So $n(108)$ is the number of all positive divisors of 108. And the same is true for $n(32)$, too.
So next, we may want to find $n(108)$ and $n(32)$. How to find them?

Factorizing first, 108 and 32, we get: $108 = 2^2 3^3$, and $32 = 2^5$.
Next, we have a fact below:

Suppose N is an integer, and $N = A^\alpha B^\beta C^\gamma \ldots Z^\delta$ where A, B, $C \ldots Z$ are primes, and α, β, γ, $\ldots \delta$ are positive integers.
Then, the number of all positive divisors of N is $(\alpha + 1)(\beta + 1)(\gamma + 1)\ldots(\delta + 1)$.

So we get: $n(108) = (2 + 1)(3 + 1) = 12$, and $n(32) = 5 + 1 = 6$.
Thus we get: $n(108)n(32)n(x) = 12 \cdot 6 \cdot n(x) = 72\ n(x) = 5400 \Rightarrow n(x) = 75$.
So we can see that 75 is the number of all positive divisors of x.
What then, can we do with the number of such divisors?

We know that the number of all positive divisors of N is $(\alpha + 1)(\beta + 1)(\gamma + 1)...(\delta + 1)$, and that α, β, γ, ... δ are positive integers, and more specifically, are the degrees of the prime factors that N has.

And we know $(\alpha + 1)(\beta + 1)(\gamma + 1)...(\delta + 1)$ is a product of integers.
So we may want to try putting 75 in a product of integers in all possible ways.

$75 = (74 + 1) \cdot 1 = (74 + 1)(0 + 1) = 25 \cdot 3 = (24 + 1)(2 + 1) = 15 \cdot 5 = (14 + 1)(4 + 1)$
$= 5 \cdot 5 \cdot 3 = (4 + 1)(4 + 1)(2 + 1)$.

Next, we want find the minimum of x.
So we want to find the smallest of all the products below:
$2^{74}3^{0} = 2^{74}1 = 2^{74}$, $2^{24}3^{2}$, $2^{14}3^{4}$, and $2^{4}3^{4}5^{2}$.

Of the three above, the smallest is $2^{4}3^{4}5^{2} = 32400$, which is thus, the minimum of x.

3.   Assuming a and b are positive integers, find the smallest a and b for which we get: $a^{2} = 2b^{2}$.

We can put $a^{2} = 2b^{2}$ this way, too: $a^{2} = b^{2} + b^{2}$. What then, can we say about a and b?

The equality looks like the distance formula.
So it seems that a is the hypotenuse in a right triangle isosceles, and b is each leg.

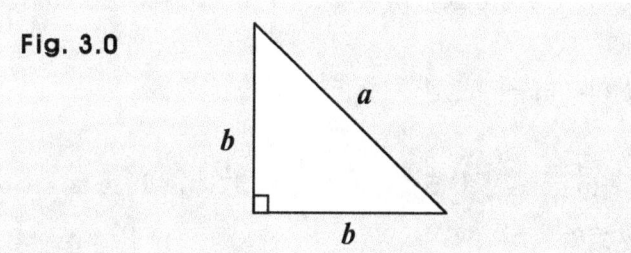

Fig. 3.0

Is there such a right triangle though? In other words, can we get a right triangle where two sides are equal and all the three sides are integers?

There is no right triangle isosceles where three sides are integers. How come?

No matter what right triangle it may be, its three sides have to satisfy the distance formula. So in the case above, too, we have to get: $a^2 = b^2 + b^2$, which is: $a^2 = 2b^2$, which is however, impossible if a and b are nonzero integers.

That is to say that if $a^2 = 2b^2$, a cannot be an integer if b is a nonzero integer. And vice versa. So if $a^2 = 2b^2$, b cannot be an integer if a is a nonzero integer.

And thus, we cannot get: $a^2 = 2b^2$ if a and b are nonzero integers.

And we can show the fact above the way below, too:

We know a and b are nonzero integers. So either prime or composite, we can put each in a form of a product of primes the way below:

Assuming first, m, n, k ... are nonnegative integers, we can set: $a = 2^m 3^n 5^k$ ....
So we get: $a^2 = (2^m 3^n 5^k ...)^2 = 2^{2m} 3^{2n} 5^{2k}$ ....

Assuming next, p, q, r ... are nonnegative integers, we can set: $b = 2^p 3^q 5^r$ ....
So we get: $b^2 = (2^p 3^q 5^r ...)^2 = 2^{2p} 3^{2q} 5^{2r}$ ..., and we get: $2b^2 = 2 \cdot 2^{2p} 3^{2q} 5^{2r} ... = 2^{2p+1} 3^{2q} 5^{2r}$ ...

If two integers are the same, their factorizations have to be the same, too.

That is to say that if $a^2 = 2b^2$, we have to get: $2^{2m} 3^{2n} 5^{2k} ... = 2^{2p+1} 3^{2q} 5^{2r}$ ..., which is however, impossible even if $m = p, n = q, k = r$, etc.

Examples Z

0. Assuming x, y, and z are integers, find the smallest y for which we get: $x^2 + z^2 = y^3$ if $x > y > z > 0$.

1. Put 2003 in the sum of powers of 2. Putting for instance, 83 in the sum of powers of 2, we get: $83 = 2^6 + 2^4 + 2^1 + 2^0$.

2. Put each of 135 and 598 in terms of the powers of digits in the number itself. For instance, $153 = 1 + 5^3 + 3^3$, $24 = 2^3 + 4^2$, and $2427 = 2 + 4^2 + 2^3 + 7^4$.

3. Show that if x is a positive integer and has an odd number of divisors, there is an integer n for which we get: $x = n^2$.

Suggestions or Solutions
To the Examples Y

0.   Assuming x, y, and z are integers, find the smallest y for which $x^2 + z^2 = y^3$ when $x > y > z > 0$.

Sometimes, trial-and-error works. It's not just random though, and is still mathematical. So it still has to make sense. So first, what do we mean by $x^2 + z^2 = y^3$?

We know x, y, and z are integers. So we can try finding an integer cubed that can be the sum of two integers squared. And next, what do we mean by $x > y > z > 0$?

The integer to be cubed is between the two integers to be squared.
So we can try some integers the way below:

$y = 1 \Rightarrow y^3 = 1, y = 2 \Rightarrow y^3 = 8, y = 3 \Rightarrow y^3 = 27 = 9 + 18.$

$y = 4 \Rightarrow y^3 = 64 = 9 + 55 = 25 + 39 = 36 + 28.$

$y = 5 \Rightarrow y^3 = 125 = 1 + 124 = 4 + 121 = 2^2 + 11^2.$ And we have: $2 < 5 < 11.$

So we can say that $x = 11, y = 5$, and $z = 2.$

1.   Put 2003 in the sum of powers of 2. Putting for instance, 83 in the sum of powers of 2, we get: $83 = 2^6 + 2^4 + 2^1 + 2^0$.

Expressing a number in terms of powers of 10, we put the number in the decimal number system where the base is 10. So we can put 2003 this way: $2 \cdot 10^3 + 3 \cdot 10^0$.
And more specifically, we can put it this way, too: $2003 = 2 \cdot 10^3 + 0 \cdot 10^2 + 0 \cdot 10^1 + 3 \cdot 10^0$.

What then, about $83 = 2^6 + 2^4 + 2^1 + 2^0$?

We can put it this way, too: $83 = 1 \cdot 2^6 + 0 \cdot 2^5 + 1 \cdot 2^4 + 0 \cdot 2^3 + 0 \cdot 2^2 + 1 \cdot 2^1 + 1 \cdot 2^0$.

And we can put $1 \cdot 2^6 + 0 \cdot 2^5 + 1 \cdot 2^4 + 0 \cdot 2^3 + 0 \cdot 2^2 + 1 \cdot 2^1 + 1 \cdot 2^0$ this way, too: 1010011.

And changing 83 to 1010011, we put 38 in the binary number system. In other words, putting 83 in the binary system, we get 1010011, which is thus, called a binary number.

2003

$= 2002 + 1$

$= 2 \cdot 1001 + 1$

$= 2(1000 + 1) + 1$

$= 2 \cdot 1000 + 2 + 1$

$= 2 \cdot (2 \cdot 500) + 2 + 1$

$= 2^2 \cdot 500 + 2 + 1$

$= 2^3 \cdot 250 + 2 + 1$

$= 2^4 \cdot 125 + 2 + 1$

$= 2^4 (124 + 1) + 2 + 1$

$= 2^4 \cdot 124 + 2^4 + 2 + 1$

$= 2^5 \cdot 62 + 2^4 + 2 + 1$

$= 2^6 \cdot 31 + 2^4 + 2 + 1$

$= 2^6 (30 + 1) + 2^4 + 2 + 1$

$= 2^6 \cdot 30 + 2^6 + 2^4 + 2 + 1$

$= 2^7 \cdot 15 + 2^6 + 2^4 + 2 + 1$

$= 2^7 (14 + 1) + 2^6 + 2^4 + 2 + 1$

$= 2^7 \cdot 14 + 2^7 + 2^6 + 2^4 + 2 + 1$

$= 2^8 \cdot 7 + 2^7 + 2^6 + 2^4 + 2 + 1$

$= 2^8 (6 + 1) + 2^7 + 2^6 + 2^4 + 2 + 1$

$= 2^8 \cdot 6 + 2^8 + 2^7 + 2^6 + 2^4 + 2 + 1$

$= 2^9 \cdot 3 + 2^8 + 2^7 + 2^6 + 2^4 + 2 + 1$

$= 2^9 (2 + 1) + 2^8 + 2^7 + 2^6 + 2^4 + 2 + 1$

$= 2^{10} + 2^9 + 2^8 + 2^7 + 2^6 + 2^4 + 2 + 1$

So we can get: $2003 = 2^{10} + 2^9 + 2^8 + 2^7 + 2^6 + 2^4 + 2 + 2^0$.

And putting 2003 in the binary system, we get: 11111010011.

And we can say that the binary equivalent to 2003 is 11111010011.

We can simplify the conversion process the way below.

First, divide 2003 by 2, and put the quotient and the remainder the way below.

2 | 2003
_____
   1001 ... 1

Next, divide 1001 by 2, and put the quotient and the remainder the way below.

2 | 2003
2 | 1001 ... 1
_____
   500 ... 1

Next, divide 1001 by 2, and put the quotient and the remainder the way below.

2 | 2003
2 | 1001 ... 1
2 | 500 ... 1
_____
   250 ... 0

And keep doing the same until the quotient is less than 2.

2 | 2003
2 | 1001 ... 1 ↑
2 | 500 ... 1
2 | 250 ... 0
2 | 125 ... 0
2 | 62 ... 1
2 | 31 ... 0
2 | 15 ... 1
2 | 7 ... 1
2 | 3 ... 1
     1 ... 1

Then, collecting the numbers in the direction of the arrow above, we get: 11111010011, which is the binary equivalent to 2003, and can be put more specifically: $11111010011_{[2]}$.

Let's now, put 2003 in the octal system where we use octal numbers.

$$
\begin{array}{r|l}
8 & 2003 \\
\hline
8 & 250 \ldots 3 \\
\hline
8 & 31 \ldots 2 \\
\hline
 & 3 \ldots 7
\end{array}
$$

That is to say that we get: $2003 = 3 \cdot 8^3 + 7 \cdot 8^2 + 2 \cdot 8 + 3 \cdot 8^0$. More specifically:

2003

$= 2000 + 3$

$= 8 \cdot 250 + 3$

$= 8(248 + 2) + 3$

$= 8 \cdot 248 + 2 \cdot 8 + 3$

$= 8 \cdot 8 \cdot 31 + 2 \cdot 8 + 3$

$= 8^2 31 + 2 \cdot 8 + 3$

$= 8^2(24 + 7) + 2 \cdot 8 + 3$

$= 8^2 24 + 7 \cdot 8^2 + 2 \cdot 8 + 3$

$= 8^3 3 + 7 \cdot 8^2 + 2 \cdot 8 + 3$

$= 3 \cdot 8^3 + 7 \cdot 8^2 + 2 \cdot 8 + 3$

$= 3723_{[8]}$

So the octal equivalent to 2003 is 3723, and more specifically, it is: $3723_{[8]}$.

2.   Put each of 135 and 598 in terms of the powers of digits in the number itself. For instance, $153 = 1 + 5^3 + 3^3$, $24 = 2^3 + 4^2$, and $2427 = 2 + 4^2 + 2^3 + 7^4$.

This problem can look difficult, but is not.

This is in particular, for those students who enjoy mental math a lot.

Sometimes, trial-and-error works. It's not random though.

So assuming first, m and n are integers, we can set: $135 = 1 + 3^m + 5^n = 1 + 9 + 125$.

So we can take 2 as m, and take 3 as n.

And thus, we get: $135 = 1 + 3^2 + 5^3$.

And assuming next, m, n, and k are integers, we can set: $598 = 5^m + 9^n + 8^k$.

$5^m = 5$, 25, and 125 if $m = 1$, 2, and 3. And m cannot be 4 or larger, because $5^4 = 625$.

$9^n = 9$ and 81 if $n = 1$ and 2. And n cannot be 3 or larger, because $9^3 = 729$.

$8^k = 8$, 64, and 512 if $k = 1$, 2, and 3. And k cannot be 4 or larger, because $8^4 = 4096$.

And finding the numbers that can make 598, we can get:

$598 = 512 + 81 + 5 = 5^1 + 9^2 + 8^3$.

3.   Show that if x is a positive integer and has an odd number of divisors, there is an integer n for which we get: $x = n^2$.

To begin with, we have a fact below:

Suppose N is an integer, and $N = A^{\alpha}B^{\beta}C^{\gamma} \dots Z^{\delta}$ where A, B, $C \dots Z$ are primes, and α, β, γ, $\dots \delta$ are positive integers.

Then, the number of all positive divisors of N is $(\alpha + 1)(\beta + 1)(\gamma + 1)\dots(\delta + 1)$.

So for instance:

$12 = 2^2 3$ has 1, 2, 3, 4, 6, and 12 as divisors, and thus, has 6 divisors.

$16 = 2^4$ has as divisors 1, 2, 4, 8, and 16, and thus, has 5 divisors.

$180 = 2^2 3^2 5$ has as divisors the ones below:

1, 2, 3, 4, 5, 6, 9, 10, 12, 15, 18, 20, 30, 36, 45, 60, 90, and 180

So it has 18 divisors.

Assuming now, m, n, $k \dots$ are integers, we can set: $x = 2^m 3^n 5^k \dots$

Then, the number of the divisors of x is: $(m + 1)(n + 1)(k + 1)\dots$, and is odd.

So we can say that every one of $m + 1$, $n + 1$, $k + 1$, etc. is odd.

That is to say that all the exponents, m, n, k, etc. are all even.

So assuming p, q, r, $\dots$ are integers, we can set: $m = 2p$, $n = 2q$, $k = 2r$, etc.

And we can get: $x = 2^m 3^n 5^k \dots = 2^{2p} 3^{2q} 5^{2r} \dots = (2^p 3^q 5^r \dots)^2$, which is an integer squared.

Sense of Arithmetic 3

0. Fill in the blanks below using no calculator.

A.

$1 + \_ = 11$ $3 + \_ = 13$ $5 + \_ = 15$ $\_ + 10 = 12$

$4 + \_ = 14$ $\_ + 10 = 17$ $\_ + 10 = 18$ $6 + \_ = 16$

$\_ + 10 = 19$ $15 = \_ + 10$ $11 = \_ + 10$ $14 = \_ + 10$

$18 = 8 + \_$ $12 = 2 + \_$ $13 = \_ + 10$ $19 = 9 + \_$

$16 = \_ + 10$ $17 = 7 + \_$

B.

1 + 10 + _ = 111   _ + 10 + 100 = 115   7 + _ + 100 = 117

2 + _ + 100 = 122   8 + 10 + _ = 318   _ + 50 + 900 = 953

6 + 70 + _ = 876   _ + 40 + 600 = 649   4 + _ + 500 = 534

C.

1 + 2 + 20 + 30 + 100 + 300 = _

2 + 4 + 40 + 10 + 200 + 500 = _

5 + 3 + 1 + 50 + 20 + 10 + 200 + 100 + 400 = _

1 + 4 + 2 + 30 + 20 + 40 + 300 + 400 + 200 = _

D.   4 + 2 + 1 + 3 + 10 + 20 + 40 + 30 + 300 + 100 + 400 + 200 = _

E.   200 + 30 + 60000 + 5 + 800000 + 4000 + 9000000 = _

1.   Find all possible and different pairs of integers nonnegative so that the sum of the two in each pair is 5, and specifying each pair, put the two integers in parentheses, and use a comma to separate the two. For example, one of the pairs is (2, 3) since $2 + 3 = 5$. Note however, that (2, 3) and (3, 2) are the same.

2.   Find all possible and different pairs of integers nonnegative so that the sum of the two in each pair is 10, and specify each pair the way above. For example, one of the pairs is (2, 8) since $2 + 8 = 10$. And of course, (2, 8) and (8, 2) are the same.

3.   In each case below, fill in the blanks with integers nonnegative so that each case is different.

A.

_ + _ = 3,   _ + _ = 3,   _ + _ = 3,   and   _ + _ + _ = 3

What other cases can make 3 each?

B.

_ + _ = 7,   _ + _ = 7,   _ + _ = 7,   and   _ + _ + _ = 7

What other cases can make 7 each?

C.

_ + _ = 6,   _ + _ = 6,   _ + _ = 6,   and   _ + _ + _ = 6

What else?

D.

$\_ + \_ = 4,$ $\_ + \_ = 4,$ $\_ + \_ = 4,$ and $\_ + \_ + \_ = 4.$ What others?

E.

$\_ + \_ = 2,$ $\_ + \_ = 2,$ $\_ + \_ = 2,$ and $\_ + \_ + \_ = 2.$ What others can do it?

F.

$\_ + \_ = 8,$ $\_ + \_ = 8,$ $\_ + \_ = 8,$ and $\_ + \_ + \_ = 8.$

There are some others. What are they?

G.

$\_ + \_ = 9,$ $\_ + \_ = 9,$ $\_ + \_ = 9,$ and $\_ + \_ + \_ = 9.$ What are others?

H.

$\_ + \_ = 11,$ $\_ + \_ = 11,$ $\_ + \_ = 11,$ $\_ + \_ + \_ = 11,$

$\_ + \_ + \_ + \_ = 11,$ and $\_ + \_ + \_ + \_ + \_ = 11.$ What else can it be?

I.

$\_ + \_ = 17,$ $\_ + \_ = 17,$ $\_ + \_ = 17,$ $\_ + \_ + \_ = 17,$

$\_ + \_ + \_ + \_ = 17,$ and $\_ + \_ + \_ + \_ + \_ = 17$

What others will do it?

J.

_ + _ = 19,   _ + _ = 19,   _ + _ = 19,   _ + _ + _ = 19,

_ + _ + _ + _ = 19,   and   _ + _ + _ + _ + _ = 19. And what else?

K.

_ + _ = 28,   _ + _ = 28,   _ + _ = 28,   _ + _ + _ = 28,

_ + _ + _ = 28,   and   _ + _ + _ + _ = 28.   And if any others, what are they?

L.

_ + _ = 39,   _ + _ = 39,   _ + _ = 39,   _ + _ + _ = 39,

_ + _ + _ = 39,   and   _ + _ + _ + _ = 39.   What else?

M.

_ + _ = 120,   _ + _ = 120,   _ + _ = 120,   _ + _ + _ = 120,

_ + _ + _ = 120,   and   _ + _ + _ + _ = 120.   What else?

N.

_ + _ = 1358, _ + _ = 1358, _ + _ = 1358,   _ + _ + _ = 1358,

_ + _ + _ = 1358,   and   _ + _ + _ + _ = 1358.   And think of some more.

118

O.

_ + _ = 98375,   _ + _ = 98375,   _ + _ = 98375,

_ + _ + _ + _ = 98375,   and   _ + _ + _ + _ + _ = 98375.   And what else can do it?

4.   Doing problems below, use nonnegative numbers.

A.

_ + _ = 2.4,   _ + _ = 2.4,   _ + _ = 2.4,   _ + _ + _ = 2.4,

_ + _ + _ + _ = 2.4,   and   _ + _ + _ + _ + _ = 2.4.   And what else?

B.

_ + _ = 0.5,   _ + _ = 0.5,   _ + _ = 0.5,   _ + _ + _ = 0.5,

_ + _ + _ + _ = 0.5,   and   _ + _ + _ + _ + _ = 0.5.   And what else?

C.

_ + _ = 0.01, _ + _ = 0.01, _ + _ = 0.01, _ + _ + _ = 0.01,

_ + _ + _ + _ = 0.01,   and,   _ + _ + _ + _ + _ = 0.01.   What others can do it?

D.

_ + _ = 0.0034,   _ + _ = 0.0034,   _ + _ = 0.0034,

_ + _ + _ = 0.0034,   _ + _ + _ + _ = 0.0034, and

_ + _ + _ + _ + _ = 0.0034.   Anything else?

E.

_ + _ = 0.9738,   _ + _ = 0.9738,   _ + _ = 0.9738,

_ + _ + _ = 0.9738,   _ + _ + _ + _ = 0.9738, and

_ + _ + _ + _ + _ = 0.9738.   And any others?

5.   Fill in the blanks with appropriate numbers.

A.

1 + _ = 1.9   _ + 0.4 = 3.4   2 + _ = 2.6   _ + 0.2 = 7.2

B.

_ + 2 + 0.4 = 32.4   10 + _ + 0.8 = 19.8   70 + 1 + _ = 71.04

C.

_ + 40 + 6 + 0.002 = 846.002   300 + 70 + 4 + _ = 374.0008

D.

_ + 5 + 300 + 80 + 0.07 = 385.27   30 + _ + 2 + 0.09 = 32.19

E.

9000 + 3 + 0.008 + _ = 9003.508   400 + 7 + 0.6 + _ = 457.6

6. Compute the following expressions with no calculator.

A. $102 + 30 =$

B. $3020 + 905 =$

C. $40081020 + 7900305 =$

D. $0.1003 + 0.02407 =$

E. $0.7020301 + 0.01050908 =$

F. $30 + 0.02 + 20 + 0.04 + 0.01 + 40 + 0.03 + 10 =$

G. $90700308002003.00690892010045 + 8024090370920.380070006012 =$

Sense of Arithmetic 4

Read the calculations below if you want to. And if you read them, just keep reading them at your pace. It can help increase sense of arithmetic.

You don't have to read them all at once. Each reading can take 5, 10, or 15 minutes at a time, or as much as you can concentrate. And having read them all, read them again when you feel like it. **Note however**, you don't have to do calculations the way below. Each calculation just shows that **there are many ways we can calculate**.

$0 = 1 \times 0 = 0 \times 1 = 0 + 0 = 2 \times 0 = 0 \times 2 = 0 + 0 + 0 = 3 \times 0 = 0 \times 3$
$= 987 \times 0 = 0 \times 987 = 0 + 0 + 0 + 0 + \ldots = 0$

$1 = 1 \times 1, 0.1 = 0.1 \times 1 = 1 \times 0.1,$ and $0.2 = 0.1 + 0.1 = 0.1 \times 2 = 2 \times 0.1$

0.3
$= 0.1 + 0.1 + 0.1 = 0.1 \times 3 = 3 \times 0.1$
$= 0.1 + 0.2 = 0.1 + (0.1 + 0.1) = 0.1 + 0.1 \times 2 = 0.1 + 2 \times 0.1$
$= 0.2 + 0.1 = (0.1 + 0.1) + 0.1 = 0.1 \times 2 + 0.1 = 2 \times 0.1 + 0.1$
$= 0 + 0.3$

0.4
$= 0.1 + 0.1 + 0.1 + 0.1 = 0.1 \times 4$
$= 0.1 + 0.3 = 0.1 + 0.1 \times 3$
$= 0.3 + 0.1 = 0.1 \times 3 + 0.1$
$= 0.2 + 0.2 = 0.2 \times 2$

0.5
$= 0.1 + 0.1 + 0.1 + 0.1 + 0.1 = 0.1 \times 5 = 5 \times 0.1$
$= 0.1 + 0.4 = 0.4 + 0.1 = 0.2 + 0.3 = 0.3 + 0.2$

122

0.6
= 0.1 x 6 = 6 x 0.1
= 0.1 + 0.5 = 0.5 + 0.1
= 0.2 + 0.4 = 0.2 + 0.2 x 2
= 0.4 + 0.2 = 0.2 x 2 + 0.2
= (0.2 + 0.2) + 0.2 = 0.2 + 0.2 + 0.2 = 0.2 x 3 = 3 x 0.2

0.7
= 0.1 x 7 = 7 x 0.1
= 0.2 + 0.5 = 0.5 + 0.2
= 0.1 + 0.6 = 0.1 + 0.2 x 3 = 0.1 + 3 x 0.2
= 0.3 + 0.4 = 0.3 + 0.2 x 2 = 0.1 x 3 + 1 x 0.2 x 2 = 1 x 0.1 x 3 + 1 x 0.2 x 2 x 1

0.8
= 8 x 0.1
= 0.1 + 0.1 + 0.1 + . . . + 0.1 = 0.1 x 8
= 0.2 + 0.2 + 0.2 + 0.2 = 0.2 x 4
= 0.4 + 0.4 = 0.4 x 2
= 0.3 + 0.5 = 0.5 + 0.3

0.9
= 0.9 x 1
= 0.1 + 0.1 + 0.1 + . . . + 0.1 = 0.1 x 9
= 0.4 + 0.5 = 0.5 + 0.4
= 0.4 + (0.4 + 0.1) = 0.4 + 0.4 + 0.1 = 0.4 x 2 + 0.1 = 2 x 0.4 + 0.1
= 0.8 + 0.1 = **2 x 0.4 + 0.1 = 4 x 0.2 + 0.1**
= 0.3 + 0.3 + 0.3 = 0.3 x 3
= 0.3 + 0.6
= 0.3 + (0.3 + 0.3) = 0.3 + 0.3 x 2 = 0.3 x 1 + 0.3 x 2 = 0.3 x (1 + 2) = 0.3 x 3
= 0.6 + 0.3
= (0.3 + 0.3) + 0.3 = 0.3 x 2 + 0.3 = 0.3 x 2 + 0.3 x 1 = 0.3 x (2 + 1) = 0.3 x 3
= 0.2 + 0.7 = 0.7 + 0.2
= 0.2 + (0.2 + 0.5)

= 0.2 + 0.2 + 0.5

= 0.2 x 2 + 0.5

= 0.2 x 2 + (0.2 + 0.3)

= 0.2 x 2 + 0.2 + 0.3

=0. 2 x 2 + 0.2 x 1 + 0.3

= 0.2 x (2 + 1) + 0.3

= **0.2 x 3 + 0.3**

= **0.3 x 2 + 0.3 x 1**

= 0.3 x (2 + 1)

= 0.3 x 3

= (0.2 + 0.1) x (2 + 1)

= (0.2 + 0.1) x 2 + (0.2 + 0.1) x 1

= (0.2 x 2 + 0.1 x 2) + (0.2 x 1 + 0.1 x 1)

= 0.2 x 2 + 0.2 + 0.2 + 0.1

= 0.4 + 0.2 + 0.2 + 0.1

= 0.4 + 0.2 x 2 + 0.1

= 0.4 + 0.4 + 0.1

= 0.4 x 2 + 0.1

= 0.8 + 0.1

1

= 10 x 0.1 + 1 x 0

= 0.1 + 0.9 = 0.9 + 0.1

= 0.2 + 0.8 = 0.8 + 0.2 = 0.8 + 0.1 + 0.1 = 0.8 + 0.1 x 2

= 0.3 + 0.7 = 0.7 + 0.3 = 0.7 + 0.2 + 0.1

= 0.4 + 0.6 = 0.6 + 0.4 = 0.6 + 0.2 + 0.2 = 0.6 + 0.2 x 2

= 0.5 + 0.5 = **0.5 x 2 = 0.2 x 5**

= 0.1 + 0.9 = 0.1 x 1 + 0.9 x 1 = 1 x 0.1 + 0.1 x 9

= 0.1 + 0.1 + 0.8 = 0.1 x 2 + 0.8 x 1 = 2 x 0.1 + 0.1 x 8

= 0.1 + 0.1 + 0.1 + 0.7 = 0.1 x 3 + 0.7 x 1 = 0.3 x 1 + 0.1 x 7 = 0.1 + 0.2 + 0.7

= 0.1 + 0.1 + 0.1 + 0.1 + 0.6

= 0.2 + 0.2 + 0.6

= 0.2 x 2 + 0.3 + 0.3

= 0.3 x 2 + 2 x 0.2 = 0.2 x 3 + 0.2 x 2 = 0.2 x (3 + 2) = 0.2 x 5

= 0.1 + 0.1 + 0.1 + 0.1 + 0.1 + 0.5

= 0.1 x 5 + 0.5 x 1

= 5 x 0.1 + 5 x 0.1

= 5 x (0.1 + 0.1)

= 5 x 0.2 = 2 x 0.5

= (0.1 + 0.1) + (0.1 + 0.1) + (0.1 + 0.1) + (0.1 + 0.1) + (0.1 + 0.1)

= 0.2 + 0.2 + 0.2 + 0.2 + 0.2

= 0.2 x 5

= 0.5 x 2

= (0.1 + 0.1 + 0.1 + 0.1 + 0.1) + (0.1 + 0.1 + 0.1 + 0.1 + 0.1)

= 0.5 + 0.5

= 5 x 0.1 + 5 x 0.1

= 2 x 5 x 0.1

= 10 x 0.1

= 0.1 + 0.1 + 0.1 + 0.1 + 0.1 + 0.1 + 0.1 + 0.1 + 0.1 + 0.1

= 0.1 + (0.1 + 0.1) + (0.1 + 0.1 + 0.1) + (0.1 + 0.1 + 0.1 + 0.1)

= 0.1 + 0.2 + 0.3 + 0.4

= (0.1 + 0.2) + (0.3 + 0.4)

= 0.3 + 0.7

= 0.1 x 3 + 0.1 x 7

= 0.1 x (3 + 7)

= 0.1 X 10

= 0.1 + 0.2 + 0.3 + 0.4

= 0.2 + 0.4 + 0.1 + 0.3

= 0.2 + 0.1 x (4 + 1) + 0.3

= 0.2 + 0.5 + 0.3 = 0.2 + 0.1 x (5 + 3) = 0.2 + 0.8

1.1
$= 1 + 0.1 = 0.5 \times 2 + 0.1 = 0.5 + 0.5 + 0.1$
$= 0.1 \times 5 + 0.1 \times 5 + 0.1$
$= 0.1 \times (5 + 5 + 1)$
$= 0.1 \times 11$

1.2
$= 1 + 0.2$
$= 0.5 \times 2 + 0.2 = 0.2 \times 5 + 0.2 \times 1 = 0.2 \times (5 + 1) = 0.2 \times 6 = 0.6 \times 2 = 0.6 + 0.6$
$= 0.4 + 0.2 + 0.4 + 0.2 = 0.4 \times 2 + 0.2 + 2 = 0.4 \times 2 + 0.4 = 0.4 \times (2 + 1)$
$= 0.4 \times 3 = 0.1 \times 4 \times 3 = 0.1 \times 3 \times 4$
$= 0.3 \times 4$

1.3
$= 1 + 0.3$
$= 0.5 \times 2 + 0.3$
$= 5 \times 0.2 + 0.2 + 0.1$
$= (5 + 1) \times 0.2 + 0.1$
$= 6 \times 0.2 + 0.1$
$= 1.2 + 0.1$
$= 0.1 \times 12 + 0.1 \times 1$
$= 0.1 \times (12 + 1)$
$= 0.1 \times 13$

1.4
$= 1 + 0.4$
$= 0.5 \times 2 + 0.4$
$= 0.2 \times 5 + 0.2 \times 2 = 0.2 \times (5 + 2) = 0.2 \times 7 = 0.7 \times 2 = 0.1 \times 7 \times 2 = 0.1 \times 14$

1.5
= 1 + 0.5
= 0.5 x 2 + 0.5
= 0.5 x (2 + 1)
= 0.5 x 3
= 0.1 x 5 x 3 = 0.1 x 15 = 15 x 0.1

1.6
= 1 + 0.6
= 1 + 0.5 + 0.1 = 0.5 x 2 + 0.5 + 0.1
= 0.5 + 0.5 + 0.5 + 0.1
= 0.5 + 0.1 + 0.4 + 0.5 + 0.1
= 0.6 + 0.4 + 0.6
= 0.6 x 2 + 0.4
= 0.1 x 6 x 2 + 0.4
= 0.1 x 12 + 0.4
= 1.2 + 0.4
= (1 + 0.2) + 0.4
= 1 + 0.2 + 0.4
= 0.5 + 0.5 + 0.2 + 0.4
= 0.4 + 0.1 + 0.4 + 0.1 + 0.2 + 0.4
= 0.4 + 0.4 + 0.4 + 0.1 + 0.1 + 0.2 = 0.4 + 0.4 + 0.4 + 0.4 = 0.4 x 4
= 0.4 x 3 + 0.1 + 0.1 + 0.2
= 0.4 x 3 + 0.4
= 0.4 x (3 + 1)
= 0.4 x (1 + 2 + 1)
= 0.4 x 1 + 0.4 x 2 + 0.4 x 1
= 0.4 + 0.8 + 0.4 = 0.4 x 2 + 0.8
= 0.8 + 0.8
= 0.8 x 2 = 0.1 x 8 x 2 = 0.1 x 2 x 8
= 0.2 x 8
= 0.1 x 2 x 8
= 0.1 x 16

$= 0.1 \times (4 + 12)$

$= 0.4 + 1.2$

$= 0.4 + 0.6 + 0.6$

$= 0.2 + 0.2 + 0.6 + 0.6$

$= 0.2 \times 2 + 0.6 \times 2 = 2 \times 0.2 + 6 \times 0.2 = 8 \times 0.2$

$= 0.2 \times 8$

$= 0.2 \times (7 + 1) = 0.1 \times 2 \times 7 + 0.1 \times 2 = 0.1 \times (14 + 2)$

1.7

$= 1 + 0.7$

$= 0.5 \times 2 + 0.7$

$= 0.5 \times 2 + (0.5 + 0.2) = 0.5 \times 3 + 0.2 = 0.1 \times 5 \times 3 + 0.1 \times 2 = 0.1 \times (15 + 2)$

1.8

$= 1 + 0.8 = 0.5 \times 2 + 0.8 = 0.5 \times 2 + 0.4 \times 2 = (0.5 + 0.4) \times 2 = 0.9 \times 2 = 2 \times 0.9$

$= 0.1 \times 10 + 0.8$

$= 0.1 \times (6 + 4) + 0.8 = 0.6 + 0.4 + 0.8$

$= 0.6 + 0.4 + 0.6 + 0.2$

$= 0.6 + 0.6 + 0.4 + 0.2$

$= 0.6 + 0.6 + 0.6$

$= 0.6 \times 3 = 0.3 \times 6 = 0.1 \times 3 \times 6 = 0.1 \times 3 \times 3 \times 2 = 0.1 \times 2 \times 9 = 0.2 \times 9$

1.9

$= 1 + 0.9$

$= 0.5 \times 2 + 0.5 + 0.4$

$= 0.5 \times 3 + 0.4 = 5 \times 0.3 + 0.4 = 0.1 \times 5 \times 3 + 0.1 \times 4 = 0.1 \times (15 + 4)$

$2 = 1 \times 2 = (0.5 \times 2) \times 2 = 0.5 \times 2 \times 2 = 0.5 \times 4 = 0.4 \times 5 = 0.1 \times 4 \times 5 = 0.1 \times 20$

$2.1 = 1 \times 2 + 0.1$

2.2

$= 2 + 0.2$

$= 1 \times 2 + 0.1 \times 2$

$= (1 + 0.1) \times 2$

$= 1.1 \times 2$

$= 0.1 \times 11 \times 2 = 0.1 \times 2 \times 11 = 0.2 \times 11$

$2.3 = 2 + 0.3 = 1 \times 2 + 0.1 \times 3$

$9.8 = 1 \times 9 + 0.8$

$9.9 = 1 \times 9 + 0.9$

$0.1 \times 10 = 1$, and $10 \times 0.1 = 1$.

$10 \div 10 = 1 \div 1 = 2 \div 2 = 5 \div 5 = 47 \div 47 = 7.2 \div 7.2 = 198 \div 198 = \ldots = 1$

$0.1 \times 100 = 0.1 \times 10 \times 10 = 1 \times 10 = 10$

$100 \times 0.1 = 10 \times 10 \times 0.1 = 10 \times 1 = 10$

$10 \div 1 = 10$ because 10 has 10 of 1s, that is, $10 \times 1 = 10$.
And 10 has 1 of 10s, so $1 \times 10 = 10$. And thus, $1 \times 10 = 10 \times 1 = 10$.

$100 \div 10 = 10$ because 100 has 10 of 10s, that is, $10 \times 10 = 100$.

$1000 \div 100 = 10$, for 1000 has 10 of 100s, that is, $10 \times 100 = 1000$.
And 1000 has 100 of 10s, so $100 \times 10 = 1000$. And thus, $100 \times 10 = 10 \times 100 = 1000$.

$10000 \div 1000 = 10$ because 10000 has 10 of 1000s, that is, $10 \times 1000 = 10000$.
And 10000 has 1000 of 10s, so $1000 \times 10 = 10000$. So $1000 \times 10 = 10 \times 1000 = 10000$.

$30 \div 3 = 10$ because 30 has 10 of 3s, that is, $10 \times 3 = 30$.
And 30 has 3 of 10s, so $3 \times 10 = 30$. And thus, $3 \times 10 = 10 \times 3 = 30$.

$50 \div 5 = 10$ because 50 has 10 of 5s, that is, $10 \times 5 = 50$.
And 50 has 5 of 10s, so $5 \times 10 = 50$. And thus, $5 \times 10 = 10 \times 5 = 50$.

$20 \div 2 = 10$ because 20 has 10 of 2s, that is, $10 \times 2 = 20$.
And 20 has 2 of 10s, so $2 \times 10 = 20$. And thus, $2 \times 10 = 10 \times 2 = 20$.

$170 \div 17 = 10$ because 170 has 10 of 17s, that is, $10 \times 17 = 170$.
And 170 has 17 of 10s, so $17 \times 10 = 170$. And thus, $17 \times 10 = 10 \times 17 = 170$.

$17 \times 10 = (10 + 7) \times 10 = 10 \times 10 + 7 \times 10 = 100 + 70 = 170$.

$3750 \div 375 = 10$ because 3750 has 10 of 375s, that is, $3750 = 10 \times 375$.
And 3750 has 375 of 10s, so $375 \times 10 = 3750$. And thus, $375 \times 10 = 10 \times 375 = 3750$.

$375 \times 10 = (300 + 70 + 5) \times 10 = 300 \times 10 + 70 \times 10 + 5 \times 10 = 3000 + 700 + 50 = 3750$.

$1 \div 0.1 = 10$ because 1 has 10 of 0.1s, that is, $1 = 10 \times 0.1$.
And 1 is one tenth of 10, that is, 1 is 0.1 of 10, so $0.1 \times 10 = 1$.
And thus, $10 \times 0.1 = 0.1 \times 10 = 1$.

$2 \div 0.2 = 10$ because 2 has 10 of 0.2s, that is, $2 = 10 \times 0.2$.

And 2 is two tenths of 10, that is, 2 is 0.2 of 10, so 0.2 x 10 = 2.

And thus, 0.2 x 10 = 10 x 0.2 = 2.

0.2 x 10 = (0.1 + 0.1) x 10 = 0.1 x 10 + 0.1 x 10 = 1 + 1 = 2.

5 ÷ 0.5 = 10 because 5 has 10 of 0.5s, that is, 10 x 0.5 = 5.

And 5 is five tenths of 10, that is, 5 is 0.5 of 10, so 0.5 x 10 = 5.

And thus, 0.5 x 10 = 10 x 0.5 = 5.

0.5 x 10 = 0.1 x 5 x 10 = 0.1 x 10 x 5 = 1 x 5 = 5.

19 ÷ 1.9 = 10 because 19 has 10 of 1.9s, that is, 10 x 1.9 = 19.

And 19 is one and nine tenths of 10, in other words, 19 is (1 + 0.9) of 10.

So (1 + 0.9) x 10 = 1.9 x 10 = 19. And thus, 1.9 x 10 = 10 x 1.9 = 19.

1.9 x 10 = 0.1 x 19 x 10 = 0.1 x 10 x 19 = 1 x 19 = 19.

174 ÷ 17.4 = 10 because 174 has 10 of 17.4s, that is, 10 x 17.4 = 174.

And 174 is seventeen and four tenths of 10, in other words, 174 is (17 + 0.4) of 10.

So (17 + 0.4) x 10 = 17.4 x 10 = 174. And thus, 17.4 x 10 = 10 x 17.4 = 174.

17.4 x 10 = (10 + 7 + 0.4) x 10 = 10 x 10 + 7 x 10 + 0.4 x 10

= 100 + 70 + 0.1 x 4 x 10 = 100 + 70 + 0.1 x 10 x 4 = 100 + 70 + 1 x 4 = 174.

Now, 0.1 x 100 = 0.1 x 10 x 10 = 1 x 10 = 10, 100 x 0.1 = 10 x 10 x 0.1 = 10 x 1 = 10,

and 10 ÷ 1 = 100 ÷ 10 = 1000 ÷ 100 = 10000 ÷ 1000

= 30 ÷ 3 = 50 ÷ 5 = 20 ÷ 2 = 170 ÷ 17 = 3750 ÷ 375

= 1 ÷ 0.1 = 2 ÷ 0.2 = 5 ÷ 0.5 = 19 ÷ 1.9 = 174 ÷ 17.4 = 10.

And we often use / (a slash) as a division sign instead of ÷.

So we can get:

10 / 1

= 100 / 10

= 1000 / 100

= 10000 / 1000

= 30 / 3

= 50 / 5

= 20 / 2

= 170 / 17

= 3750 / 375

= 1 / 0.1

= 2 / 0.2

= 5 / 0.5

= 19 / 1.9

= 174 / 17.4

= 10.

Now, we have: $1 \times 0.1 = 0.1$, and have: $1/10 = 0.1$ since $1/10 = 1 \div 10 = 0.1$. So we can get:

$2/10 = (1 + 1)/10 = 1/10 + 1/10 = 2 \times (1/10) = 2 \times 0.1 = 0.2$.

$3/10 = (1 + 2)/10 = 1/10 + 2/10 = 0.1 + 0.2 = 0.3$

$4/10 = (1 + 3)/10 = 1/10 + 3/10 = 0.1 + 0.3 = 0.4$

. . .

$9/10 = (1 + 8)/10 = 1/10 + 8/10 = 0.1 + 0.8 = 0.9$

And next, we have: $1/100 = 0.01$. So we can get:

$2/100 = (1 + 1)/100 = 1/100 + 1/100 = 0.01 + 0.01 = 0.02$

$3/100 = (1 + 2)/100 = 1/100 + 2/100 = 0.01 + 0.02 = 0.03$

$4/100 = (1 + 3)/100 = 1/100 + 3/100 = 0.01 + 0.03 = 0.04$

. . .

$9/100 = (1 + 8)/100 = 1/100 + 8/100 = 0.01 + 0.08 = 0.09$

And next, we have: $0.1/10 = 0.01$. So we can get:

$0.2/10 = (0.1 + 0.1)/10 = 0.1/10 + 0.1/10 = 2 \times (0.1/10) = 2 \times 0.01 = 0.02.$
$0.3/10 = (0.1 + 0.2)/10 = 0.1/10 + 0.2/10 = 0.01 + 0.02 = 0.03$
$0.4/10 = (0.1 + 0.3)/10 = 0.1/10 + 0.3/10 = 0.01 + 0.03 = 0.04$
. . .
$0.9/10 = (0.1 + 0.8)/10 = 0.1/10 + 0.8/10 = 0.01 + 0.08 = 0.09$

Now, we have: $1/10 = 0.1$, $1/100 = 0.01$, and $0.1/10 = 0.01$.
And also, we know: $0.1 \times 1 = 1 \times 0.1 = 0.1$, and $0.1 \times 10 = 10 \times 0.1 = 1$.
So we can get: $10 \times 0.1 = 10 \times (1/10) = 10/10 = 1$.

And also, $0.1 \times 0.1 = 0.1/10 = 0.01$ since $0.1/10 = 0.1 \div 10$.
And thus, $0.1 \times 0.1 = 0.01$.
So we can get: $0.01 = 0.1 \times 0.1 = (1/10) \times (1/10) = (1/10)/10$.

And next, we have: $(1/10)/10 = 1/10/10$. And we have: $0.1 = 1/10$, and $0.1/10 = 0.01$.
So we get: $1/10/10 = 0.1/10 = 0.01$.
And thus, we get: $1/10/10 = 0.01 = 1/100$. So we get: $1/10/10 = 1/100$.

What is $1/10$ though?

Dividing 1 into ten (10) equal parts, and taking one of the parts, we get: $1/10 = 0.1$.
What then, is $(1/10)/10$?

Dividing $1/10$ into ten (10) equal parts, and taking one of the parts, we get: $(1/10)/10$, which is $1/10/10$.

And dividing 0.1 into 10 equal parts, and taking one of the parts, we get: 0.01, which is 1/100.

We know: 1/10 = 0.1. So we get: 1/10/10 = 1/100.

And also, dividing 1 into a hundred equal parts, and taking one of the parts, we get 1/100. And thus, we get: (1/10)/10 = 1/10/10 = 0.1/10 = 1/100 = 0.01.

And by the same token, we can get:

1/10/10/10 = 1/100/10 = 1/1000 = 0.001,
1/10/10/10/10 = 1/10/100/10 = 1/10/1000 = 1/1000/10 = 1/10000 = 0.0001.

And again, by the same token, we can get:
3/10 = 0.3, 3/10/10 = 3/100 = 0.03, 3/10/10/10 = 3/10/100 = 3/1000 = 0.003, ...

And also, we can get:

3/2/7 = 3/14

3/2/7/5 = 3/14/5 = 3/2/35 = 3/70

10
= (0.1 + 0.9) x 10 = (0.9 + 0.1) x 10 = 9 + 1

= (0.2 + 0.8) x 10 = (0.8 + 0.2) x 10 = 8 + 1 + 1

= (0.3 + 0.7) x 10 = (0.7 + 0.3) x 10 = 7 + 2 + 1

= (0.4 + _) x 10 = (0.6 + _) x 10 = 6 + 2 + 2 = 6 + 2 x _
= (0.5 + _) x 10 = 0.5 x 2 x _ = 0.1 x 2 x 5 x _
= (0.1 + _) x 10 = 0.1 x 10 + 0.9 x _ = 1 + _ x 9

$= (0.1 + 0.1 + 0.8) \text{ x } 10 = (0.1 \text{ x } 2 + 0.8 \text{ x } 1) \text{ x } 10 = (0.2 \text{ x } 1 + 0.1 \text{ x } 8) \text{ x } 10$

$= (0.1 + 0.1 + 0.1 + 0.7) \text{ x } 10$

$= (0.1 \text{ x } 3 + 0.7 \text{ x } 1) \text{ x } 10$

$= (3 \text{ x } 0.1 + 0.1 \text{ x } 7) \text{ x } 10$

$= (0.3 \text{ x } 1 + 7 \text{ x } 0.1) \text{ x } 10 = (0.1 + 0.2 + 0.7) \text{ x } 10$

$= (0.1 + 0.1 + 0.1 + 0.1 + 0.6) \text{ x } 10$

$= (0.2 + 0.2 + 0.6) \text{ x } 10$

$= (0.2 \text{ x } 2 + 0.3 + 0.3) \text{ x } 10$

$= (0.3 \text{ x } 2 + 0.2 \text{ x } 2) \text{ x } 10$

$= (2 \text{ x } 0.3 + 2 \text{ x } 0.2) \text{ x } 10 = 2 \text{ x } (0.3 + 0.2) \text{ x } 10 = 2 \text{ x } 0.5 \text{ x } 10$

$= (0.1 + 0.1 + 0.1 + 0.1 + 0.1 + 0.5) \text{ x } 10$

$= (0.1 \text{ x } 5 + 0.5 \text{ x } 1) \text{ x } 10$

$= (5 \text{ x } 0.1 + 5 \text{ x } 0.1) \text{ x } 10 = (5 \text{ x } 0.1) \text{ x } 2 \text{ x } 10$

$= 0.5 \text{ x } (1 + 1) \text{ x } 10$

$= 0.5 \text{ x } 2 \text{ x } 10$

$= (0.5 + 0.5) \text{ x } 10$

$= \{(0.1 + 0.1 + 0.1 + 0.1 + 0.1) + (0.1 + 0.1 + 0.1 + 0.1 + 0.1)\} \text{ x } 10$

$= \{(0.2 + 0.3) + (0.1 + 0.4)\} \text{ x } 10$

$= (0.2 + 0.3) \text{ x } 10 + (0.1 + 0.4) \text{ x } 10$

$= 2 + 3 + 1 + 4 = 10$

$= 0.1 \text{ x } 10 \text{ x } 10$

$= (0.1 + 0.1 + 0.1 + 0.1 + 0.1 + 0.1 + 0.1 + 0.1 + 0.1 + 0.1) \text{ x } 10$

$= \{0.1 + (0.1 + 0.1) + (0.1 + 0.1 + 0.1) + (0.1 + 0.1 + 0.1 + 0.1)\} \text{ x } 10$

$= (0.1 + 0.2 + 0.3 + 0.4) \text{ x } 10$

$= \{(0.1 + 0.2) + (0.3 + 0.4)\} \text{ x } 10 = (0.3 + 0.7) \text{ x } 10$

$= 1 \text{ x } 10$

$= (0.1 + 0.2 + 0.3 + 0.4) \text{ x } 10$

$= (0.2 + 0.4 + 0.1 + 0.3) \text{ x } 10$

$= \{0.2 + (0.4 + 0.1) + 0.3\} \text{ x } 10$

$= (0.2 + 0.5 + 0.3) \text{ x } 10$

$= \{0.2 + (0.5 + 0.3)\} \text{ x } 10 = (0.2 + 0.8) \text{ x } 10 = 1 \text{ x } 10$

1.01
= 1 + 0.01
= 1 x 1 + 0.01 x 1
= 1 x 1 + 0.1 x 0 + 0.01 x 1

1.02
= 1 + 0.02
= 1 x 1 + 0.01 x 2
= 1 x 1 + 0.1 x 0 + 0.01 x 2

1.03
= 1 + 0.03
= 1 x 1 + 0.01 x 3
= 1 x 1 + 0.1 x 0 + 0.01 x 3
1.08
= 1 + 0.08
= 1 x 1 + 0.01 x 8
= 1 x 1 + 0.1 x 0 + 0.01 x 8

1.09
= 1 + 0.09
= 1 x 1 + 0.1 x 0 + 0.01 x 9

0.11
= 0.1 + 0.01
= 1 x 0 + 0.1 x 1 + 0.01 x 1

0.111
= 0.1 + 0.01 + 0.001
= 0.1 x 1 + 0.01 x 1 + 0.001 x 1

0.112
= 0.1 + 0.01 + 0.002
= 0.1 x 1 + 0.01 x 1 + 0.001 x 2

0.113
= 0.1 x 1 + 0.01 x 1 + 0.001 x 3
= 1 x 0.1 + 1 x 0.01 + 3 x 0.001

0.114
= 0.1 + 0.01 + 0.004
= 1 x 0.1 + 1 x 0.01 + 4 x 0.001

0.117
= 0.1 + 0.01 + 0.007
= 1 x 0.1 + 1 x 0.01 + 7 x 0.001

0.127
= 0.1 + 0.02 + 0.007
= 0.1 x 1 + 0.01 x 2 + 0.001 x 7
= 1 x 0.1 + 2 x 0.01 + 7 x 0.001

0.128
= 0.1 + 0.02 + 0.008
= 0.1 x 1 + 0.01 x 2 + 0.001 x 8
= 1 x 0.1 + 2 x 0.01 + 8 x 0.001

0.129
= 0.1 + 0.02 + 0.009
= 0.1 x 1 + 0.01 x 2 + 0.001 x 9
= 1 x 0.1 + 2 x 0.01 + 9 x 0.001

0.13

= 0.1 + 0.03

= 0.1 x 1 + 0.01 x 3 + 0.001 x 0

= 1 x 0.1 + 3 x 0.01 + 0 x 0.001

0.131

= 0.1 + 0.03 + 0.001

= 0.1 x 1 + 0.01 x 3 + 0.001 x 1

= 1 x 0.1 + 3 x 0.01 + 1 x 0.001

0.132

= 0.1 + 0.03 + 0.002

= 0.1 x 1 + 0.01 x 3 + 0.001 x 2

= 1 x 0.1 + 3 x 0.01 + 2 x 0.001

= 1 x 0.1 + 3 x 0.1 x 0.1 + 2 x 0.1 x 0.01

= 1 x 0.1 + 3 x 0.1 x 0.1 + 2 x 0.1 x 0.1 x 0.1

0.199

= 0.1 + 0.09 + 0.009

= 0.1 x 1 + 0.01 x 9 + 0.001 x 9

= 1 x 0.1 + 9 x 0.01 + 9 x 0.001

= 1 x 0.1 + 9 x 0.1 x 0.1 + 9 x 0.1 x 0.01

= 1 x 0.1 + 9 x 0.1 x 0.1 + 9 x 0.1 x 0.1 x 0.1

0.2

= ... + 100 x 0 + 10 x 0 + 1 x 0 + 0.1 x 2 + 0.01 x 0 + 0.001 x 0 + ...

= ... + 0 x 100 + 0 x 10 + 0 x 1 + 2 x 0.1 + 0 x 0.01 + 0 x 0.001 + ...

= ... + 0 x 10x10 + 0 x 10 + 0 x 1 + 1 x 0.1 + 0 x 0.1x0.1 + 0 x 0.1x0.1x0.1 + ...

0.201
= 0.2 + 0.001
= 0.1 x 2 + 0.01 x 0 + 0.001 x 1
= 2 x 0.1 + 0 x 0.01 + 1 x 0.001
= 2 x 0.1 + 0 x 0.1 x 0.1 + 1 x 0.1 x 0.1 x 0.1

0.248
= 0.2 + 0.04 + 0.008
= 0.1 x 2 + 0.01 x 4 + 0.001 x 8
= 2 x 0.1 + 4 x 0.01 + 8 x 0.001
= 2 x 0.1 + 4 x 0.1 x 0.1 + 8 x 0.1 x 0.1 x 0.1

0.249
= 0.2 + 0.04 + 0.009
= 0.1 x 2 + 0.01 x 4 + 0.001 x 9
= 2 x 0.1 + 4 x 0.01 + 9 x 0.001
= 2 x 0.1 + 4 x 0.1 x 0.1 + 9 x 0.1 x 0.1 x 0.1

0.794
= 0.7 + 0.09 + 0.004
= 0.1 x 7 + 0.01 x 9 + 0.001 x 4
= 7 x 0.1 + 9 x 0.01 + 4 x 0.001
= 7 x 0.1 + 9 x 0.1 x 0.1 + 4 x 0.1 x 0.1 x 0.1

0.795
= 0.7 + 0.09 + 0.005
= 0.1 x 7 + 0.01 x 9 + 0.001 x 5
= 7 x 0.1 + 9 x 0.01 + 5 x 0.001
= 7 x 0.1 + 9 x 0.1 x 0.1 + 5 x 0.1 x 0.1 x 0.1

0.796

= 0.7 + 0.09 + 0.006

= 0.1 x 7 + 0.01 x 9 + 0.001 x 6

= 7 x 0.1 + 9 x 0.01 + 6 x 0.001

= 7 x 0.1 + 9 x 0.1 x 0.1 + 6 x 0.1 x 0.1 x 0.1

0.997

= 0.9 + 0.09 + 0.007

= 0.1 x 9 + 0.01 x 9 + 0.001 x 7

= 9 x 0.1 + 9 x 0.01 + 7 x 0.001

= 9 x 0.1 + 9 x 0.1 x 0.1 + 7 x 0.1 x 0.1 x 0.1

Note that $10 = 10^1$, $0.1 = 0.1^1$, and $1/10 = 1/10^1 = (1/10)^1$.

{10, 100, 1000 . . .}

= {10, 10 x 10, 10 x 10 x 10 . . .}

= {10^1, 10^2, 10^3 . . .}

{0.1, 0.01, 0.001 . . .}

= {0.1, 0.1 x 0.1, 0.1 x 0.1 x 0.1 . . .}

= {0.1^1, 0.1^2, 0.1^3 . . .}

= {1/10, 1/100, 1/1000 . . .}

= {$1/10^1$, $1/10^2$, $1/10^3$ . . .}

= {1/10, 1/10/10, 1/10/10/10 . . .}

= {1/10, (1/10) x (1/10), (1/10) x (1/10) x (1/10) . . .}

= {$(1/10)^1$, $(1/10)^2$, $(1/10)^3$ . . .}

0.998

= 0.9 + 0.09 + 0.009

= 0.1 x 9 + 0.01 x 9 + 0.001 x 8

= 9 x 0.1 + 9 x 0.01 + 8 x 0.001

= 9 x 0.1 + 9 x 0.1 x 0.1 + 8 x 0.1 x 0.1 x 0.1

0.999

$= 0.9 + 0.09 + 0.009$

$= 0.1 \times 9 + 0.01 \times 9 + 0.001 \times 9$

$= 9 \times 0.1 + 9 \times 0.01 + 9 \times 0.001$

$= 9 \times 0.1 + 9 \times 0.1 \times 0.1 + 9 \times 0.1 \times 0.1 \times 0.1$

1

$= 0.999 + 0.001$

$= 9 \times 0.1 + 9 \times 0.1 \times 0.1 + 9 \times 0.1 \times 0.1 \times 0.1 + 1 \times 0.1 \times 0.1 \times 0.1$

$= 9 \times 0.1 + 9 \times 0.1 \times 0.1 + (9 + 1) \times 0.1 \times 0.1 \times 0.1$

$= 9 \times 0.1 + 9 \times 0.1 \times 0.1 + 10 \times 0.1 \times 0.1 \times 0.1$

$= 9 \times 0.1 + 9 \times 0.1 \times 0.1 + 1 \times 0.1 \times 0.1$

$= 9 \times 0.1 + (9 + 1) \times 0.1 \times 0.1$

$= 9 \times 0.1 + 10 \times 0.1 \times 0.1$

$= 9 \times 0.1 + 1 \times 0.1$

$= (9 + 1) \times 0.1$

$= 10 \times 0.1 = 0.1 \times 10$

$1 = 0.01 \times 100$

$= (0.001 + 0.009) \times 100 = 0.001 \times (1 + 9) \times 100 = 0.001 \times 10 \times 100 = 0.001 \times 1000$

$= (0.002 + 0.008) \times 100$

$= (0.003 + 0.007) \times 100$

$= (0.004 + 0.006) \times 100$

$= (0.005 + 0.005) \times 100$

$= 0.1 \times 10$

$= (0.01 + 0.09) \times 10 = 0.01 \times (1 + 9) \times 10 = 0.01 \times 10 \times 10 = 0.01 \times 100$

$= (0.02 + 0.08) \times 10$

$= (0.03 + 0.07) \times 10$

$= (0.04 + 0.07) \times 10$

$= (0.05 + 0.05) \times 10$

$= 0.1 + 0.9 = 0.1 \times (1+9) = 0.1 \times 10 = 0.2 + 0.8 = 0.3 + 0.7 = 0.4 + 0.6 = 0.5 + 0.5$

0.1985
= 0.1 + 0.09 + 0.008 + 0.0005
= 0.1 x 1 + 0.01 x 9 + 0.001 x 8 + 0.0001 x 5
= 1 x 0.1 + 9 x 0.01 + 8 x 0.001 + 5 x 0.0001

0.3092
= 0.3 + 0.009 + 0.0002
= 0.1 x 3 + 0.01 x 0 + 0.001 x 9 + 0.0001 x 2
= 3 x 0.1 + 9 x 0.001 + 2 x 0.0001

0.8964
= 0.8 + 0.09 + 0.006 + 0.0004
= 8 x 0.1 + 9 x 0.01 + 6 x 0.001 + 4 x 0.0001

0.8974
= 0.8 + 0.0974
= 0.8 + 0.09 + 0.0074
= 0.8 + 0.09 + 0.007 + 0.0004
= 0.89 + 0.007 + 0.0004
= 0.897 + 0.0004 = 0.807 + 0.0904
= 0.8004 + 0.097 = 0.0004 + 0.897

0.9999
= 0.9 + 0.09 + 0.009 + 0.0009
= 9 x 0.1 + 9 x 0.01 + 9 x 0.001 + 9 x 0.0001
= 9 x 0.1 + 9 x 0.1 x 0.1 + 9 x 0.1 x 0.01 + 9 x 0.1 x 0.001
= 9 x 0.1 + 9 x 0.1 x 0.1 + 9 x 0.1 x 0.1 x 0.1 + 9 x 0.1 x 0.1 x 0.01
= 9 x 0.1 + 9 x 0.1 x 0.1 + 9 x 0.1 x 0.1 x 0.1 + 9 x 0.1 x 0.1 x 0.1 x 0.1

1
$= 0.9999 + 0.0001$
$= 9 \times 0.1 + 9 \times 0.01 + 9 \times 0.001 + 9 \times 0.0001 + 0.0001$
$= 9 \times 0.1 + 9 \times 0.01 + 9 \times 0.001 + (9 + 1) \times 0.0001$
$= 9 \times 0.1 + 9 \times 0.01 + 9 \times 0.001 + 10 \times 0.0001$
$= 9 \times 0.1 + 9 \times 0.01 + 9 \times 0.001 + 10 \times 0.1 \times 0.001$
$= 9 \times 0.1 + 9 \times 0.01 + 9 \times 0.001 + 1 \times 0.001$
$= 9 \times 0.1 + 9 \times 0.01 + (9 + 1) \times 0.001$
$= 9 \times 0.1 + 9 \times 0.01 + 10 \times 0.001$
$= 9 \times 0.1 + 9 \times 0.01 + 1 \times 0.01$
$= 9 \times 0.1 + (9 + 1) \times 0.01$
$= 9 \times 0.1 + 10 \times 0.01$
$= 9 \times 0.1 + 1 \times 0.1$
$= (9 + 1) \times 0.1$
$= 10 \times 0.1$
$= 1$

6974.835
$= 6000 + 900 + 70 + 4 + 0.8 + 0.03 + 0.005$

$= 6 \times 1000 + 9 \times 100 + 7 \times 10 + 4 \times 1 + 8 \times 0.1 + 3 \times 0.01 + 5 \times 0.001$

$= 6 \times 10 \times 100 + 9 \times 10 \times 10 + 7 \times 10 + 4 \times (10/10) + 8 \times 0.1 + 3 \times 0.1 \times 0.1$
$+ 5 \times 0.1 \times 0.01$

$= 6 \times 10 \times 10 \times 10 + 9 \times 10 \times 10 + 7 \times 10 + 4 \times (10/10) + 8 \times 0.1 + 3 \times 0.1 \times 0.1$
$+ 5 \times 0.1 \times 0.1 \times 0.1$

$= 6 \times 10 \times 10 \times 10 + 9 \times 10 \times 10 + 7 \times 10 + 4 \times (10/10) + 8 \times (1/10)$
$+ 3 \times (1/10) \times (1/10) + 5 \times (1/10) \times (1/10) \times (1/10)$

$= 6 \times 1000 + 9 \times 100 + 7 \times 10 + 4 \times 1 + 8 \times (1/10) + 3 \times (1/100) + 5 \times (1/1000)$
$= 6 \times 10^3 + 9 \times 10^2 + 7 \times 10^1 + 4 \times 10^0 + 8 \times 10^{-1} + 3 \times 10^{-2} + 5 \times 10^{-3}$

www.ingramcontent.com/pod-product-compliance
Lightning Source LLC
Chambersburg PA
CBHW081452170526
45166CB00008B/2406